Palgrave Studies in Green Criminology

Series Editors

Angus Nurse, Nottingham Trent University, Nottingham, UK
Rob White, School of Social Sciences, University of Tasmania,
Hobart, TAS, Australia
Melissa Jarrell, Department of Social Sciences, Texas A&M
University - Corpus Christi, Corpus Christi, TX, USA

Criminologists have increasingly become involved and interested in environmental issues to the extent that the term Green Criminology is now recognised as a distinct subgenre of criminology. Within this unique area of scholarly activity, researchers consider not just harms to the environment, but also the links between green crimes and other forms of crime, including organised crime's movement into the illegal trade in wildlife or the links between domestic animal abuse and spousal abuse and more serious forms of offending such as serial killing. This series will provide a forum for new works and new ideas in green criminology for both academics and practitioners working in the field, with two primary aims: to provide contemporary theoretical and practice-based analysis of green criminology and environmental issues relating to the development of and enforcement of environmental laws, environmental criminality, policy relating to environmental harms and harms committed against non-human animals and situating environmental harms within the context of wider social harms; and to explore and debate new contemporary issues in green criminology including ecological, environmental and species justice concerns and the better integration of a green criminological approach within mainstream criminal justice. The series will reflect the range and depth of high-quality research and scholarship in this burgeoning area, combining contributions from established scholars wishing to explore new topics and recent entrants who are breaking new ground.

Yarin Eski

A Criminology of the Human Species

Setting an Unsettling Tone

Yarin Eski
VU University Amsterdam
Amsterdam, Noord-Holland
The Netherlands

Palgrave Studies in Green Criminology
ISBN 978-3-031-36091-6   ISBN 978-3-031-36092-3   (eBook)
https://doi.org/10.1007/978-3-031-36092-3

Cover illustration: © Melisa Hasan

This Palgrave Macmillan imprint is published by the registered company Springer Nature Switzerland AG
The registered company address is: Gewerbestrasse 11, 6330 Cham, Switzerland

For my father, Yaşar Eski

PREFACE

Whenever I listen to *Danse Macabre* by the Romantic composer Camille Saint-Saëns, the opening dissonant tones played by the violin immediately catch my imagination of the thematic of the song: a memento mori, or, how all that is alive shall eventually pass away. The dissonant tones capture in sound that unnerving, atrabilious reminder of our life toward certain death.

However, as Professor Emeritus of musicology Roger Kamien (2008) once argued, such dissonant, unstable tones form a tension that demands an onward musical motion to a stable chord. Dissonant, unsettling tones, as such, may be harsh as they express pain, conflict and death. However, they should be considered active and a way of working toward consonance and harmony. Certain death leads to certain new life.

This book, *A Criminology of the Human Species. Setting an Unsettling Tone*, brings about such dissonance. To avoid any confusion though, this book will not deliver theory or detailed empirical studies on the human species as a whole. That would be overly pretentious, given the scale of the topic. However, it aims to set an unsettling tone by *imagining* human beings as a deviant species. Such an imagination can pave the way for a profound theoretical, methodological and empirical criminology of the human species.

By delving into rather unfamiliar disciplines for criminology, this book brings together, among others, evolutionary biology, astronomy, palaeontology and posthumanism. Equipped with interdisciplinary perspectives, the reader will be taken on a historical-topical journey. From the beginning of Earth into the future of humankind, significant cases will be scrutinised in order to tune criminology into a disquieting imagination of the human species.

Criminology tends to shy away from such existential exercises that could lead to new, creative and perhaps hopeful criminologies. This book will not do so and attempts, if anything, to trigger agitation, discontent, uneasiness and overall tonal dissonance, which, if achieved, would already be more than enough.

Amsterdam, The Netherlands   Yarin Eski

CONTENTS

Abbreviations

AGI	Artificial General Intelligence
AI	Artificial Intelligence
CCR5	Cysteine-Cysteine Chemokine Receptor 5
ChatGPT	Chat Generative Pre-trained Transformer
CRISPR	Clustered Regularly Interspaced Short Palindromic Repeats
DART Mission	Double Asteroid Redirection Test Mission
ESA	European Space Agency
IGA	ISS Intergovernmental Agreement
IPCC	Intergovernmental Panel on Climate Change
ISS	International Space Station
JWST	James Webb Space Telescope
NASA	National Aeronautics and Space Administration
NEO	Near-Earth Object
OST	Outer Space Treaty
SETI	Search for Extraterrestrial Intelligence
UN	United Nations
US	United States
USSR	Union of Soviet Socialist Republics

LIST OF ILLUSTRATIONS

Dinosaurs, Hot Summers, the James Webb Telescope and the Criminological Imagination: An Introduction

Abstract This first chapter serves as an introduction to the main topics and imaginative starting point of the book, which is the criminological imagination (Young, 2011). It will become clear why it matters to work toward a criminology of the human species as a criminal, fatal and most violently destructive species—a criminology that is best served by establishing a criminological imagination of the human species.

Keywords Criminological imagination · Anthropocene · Human species

1.1   First Thoughts

When the idea for this book was conceived in 2022,[1] a couple of events had taken place during the summer of that year. In Europe, it was one of the hottest summers in recorded history, paired with unusually dry conditions. It led to record temperatures, (soil) drought and fires, including

---

[1] Parts of this book are based on an article I wrote, originally in Dutch and translated into English as: 'Omnia cadunt. Toward a victimological imagination of our transience' (Eski, 2022).

© The Author(s), under exclusive license to Springer Nature Switzerland AG 2023
Y. Eski, *A Criminology of the Human Species*,
Palgrave Studies in Green Criminology,
https://doi.org/10.1007/978-3-031-36092-3_1

the one in France where the highest levels of carbon pollution from wild-fires were measured (Vamburg in Galey, 2022). The hot summers are a consequence of a worsening climate change. The UN Intergovernmental Panel on Climate Change (IPCC) Report (2021) alarmingly concluded that humans are the main cause of unprecedented rapid and widespread climate change. CO_2 emissions have caused serious and perhaps irreversible damage to the atmosphere, oceans and polar regions. If action is not taken imminently, the planet will heat up irreversibly to such an extent that we will be doomed to live on an uninhabitable planet.

Despite such doom and gloom, the hot summer of 2022 was also one of new hopes for all humankind, as the James Webb Space Telescope's (JWST) first pictures of the universe were shared, including one of the Carina Nebula:

> This landscape of "mountains" and "valleys" speckled with glittering stars is actually the edge of a nearby, young, star-forming region called NGC 3324 in the Carina Nebula. Captured in infrared light by NASA's new James Webb Space Telescope, this image reveals for the first time previously invisible areas of star birth. Called the Cosmic Cliffs, Webb's seemingly three-dimensional picture looks like craggy mountains on a moonlit evening. In reality, it is the edge of the giant, gaseous cavity within NGC 3324, and the tallest "peaks" in this image are about 7 light-years high. The cavernous area has been carved from the nebula by the intense ultraviolet radiation and stellar winds from extremely massive, hot, young stars located in the center of the bubble, above the area shown in this image. (NASA, 2022—online source)

The pictures symbolise '[t]he dawn of a new era in astronomy [that] has begun […] to unfold the infrared universe' (id.), NASA claims. Not only does the JWST give visual insights into deep space for space exploration, it also strengthens our deeper wish and imagination to have a better future away from a climate-changed, ruined Earth (Dunnett et al., 2019; Lasch, 1979). NASA's Artemis II mission to the Moon and human settlement on Mars initiatives solidify such ambitions (NASA, 2023; Szocik, 2019).

So, across the world, the summer of 2022 was one of extremes: of desperation over earthly climate change on the one hand and of hopeful shiny celestial pictures and ambitions of space travelling on the other. For me personally it was also the summer of the latest sequel in my all-time favourite *Jurassic Park* film series, *Jurassic World: Dominion*

(Trevorrow, 2022). It has been referred to as the best example of hyperbolic Hollywood entertainment, with an underlying cautionary narrative of technological hubris that is very real (Maynard, 2022—online source). The film captures what happens when we ignore the 'dangers of unfettered entrepreneurship and irresponsible innovation', and let science and technology play God with nature's genetic code, especially by resurrecting extinct species without any consensus on what ethical and responsible progress and advancement may look like (id.). In fact, extinct species resurrection is no longer just science fiction (Brisman & South, 2020). There are initiatives that want to revive the extinct woolly mammoth (Revive & Restore, 2023) as early as 2027 (Colossal, 2023).

It made me wonder how it is possible that when it is scientifically proven that our planet is rapidly changing into a dead planet, like Mars, we simultaneously want to go to and terraform Mars into a liveable place for humankind to prosper (cf. Persinger, 2020). And at the same time, we are aiming to revive extinct species from previous ecosystems that were part of a completely different biodiversity. I asked myself: 'Who in the face of their extinction doesn't do everything and anything to prevent their extinction from happening, but rather dreams of living in extraterrestrial harsh environments and tries to resurrect extinct species on Earth? Who does that?!' We, the human species.

But why and how do we do so? As an attempt to answer that question, it is anticipated that our species' past, present and future ambitions could assist, when considered from a criminologically imaginative point of view. Accordingly, this book provides a criminological imagination of the human species that is a first step toward a criminology of the human species. However, why should *criminology* scrutinise humankind as a species? There are many other disciplines that study the human species, such as biology and palaeoanthropology, indicating that criminology would have no business gaining knowledge on the human species.

Nevertheless, criminology is more than just the study of crime and law-breaking behaviours (Ferrell & Sanders, 1995; Ferrell et al., 2004; Garland, 1992). Criminology includes the study of crime policy and policing, public perceptions of crime and victimhood, state and corporate crimes, war and human rights violations, and ecological crime and harms, on Earth and in space[2] (cf. Brisman & South, 2013, 2020; Green &

---

[2] In criminology, where spatial analyses have become commonplace, "space" has a different connotation, as it refers to the physical context of a city, university, train station

Ward, 2000; Lampkin, 2021; McGarry & Walklate, 2019; Slapper & Tombs, 1999; Tombs, 2018). Criminology is in that sense also about understanding why and how violations, including crime, elicit strong emotional responses, such as outrage and disgust from society, revealing insights into the cultural and social context, as well as the moral and ethical values of society (Boutellier, 2019). The diverse topics criminologists are concerned with makes criminology a hybrid discipline (Pease, 2021) that brings together a range of disciplines and approaches in order to understand harm, deviancy, crime, and control from a variety of perspectives. It also accommodates the study of harmful acts that remain legal (crimes that are not necessarily violations of the law), which requires a certain lens that scopes unlawful *and* lawful crime.

Especially now, such a criminological lens, or imagination, of the human species is necessary. We, the human species, are confronted with the gravest (self-inflicted) still not criminalised crime imaginable to us: mass extinction. It is the most annihilating and exploitative act the human species has managed to bring upon itself and the entire planet (Lees et al., 2020; Padilla, 2021; Pievani, 2014). And still, as we are becoming extinct, the human survival narrative still dominates the public debate (Broderick, 1993; Doyle, 2015; Norgaard, 2020). Therefore, like biology, environmental studies, religious studies and philosophy (cf. Barnosky et al., 2011; Cairns, 2013; Campbell, 1974; Chowdhury et al., 2021; Clark, 1989; Gee, 2021; Glikson, 2021; Hirschfeld & Blackmer, 2021), criminology overall should study (more) the human species and its totally annihilating and exploitative tendencies in relation to the Anthropocene mass extinction (cf. Brisman & South, 2020), human enhancement and (artificially intelligent) singularity, and human settlement in outer space. Aiming to become a giant leap for criminology, this book is a small but first step by a criminologist. It delivers a criminological imagination of the human species and the ways in which we act as a deviant, fatal, exploitative and most violent species, having done so from the beginning and will continue to do so into the future. But why a criminological imagination?

---

and so on, and its interplay with criminality (Hayward, 2016). Here, the term "space" has the meaning commonly used when referring to "outer space".

1.2 The Criminological Imagination

Logic will get you from A to B. Imagination will get you everywhere. (Albert Einstein)

Perhaps upon reading the title *A Criminology of the Human Species*, an average criminologist might wonder, 'Hang on, hasn't a criminologist already examined the criminal man before?' And indeed, this is the case. In fact, two key thinkers have already considered our species' criminality: Adolphe Quetelet and Cesare Lombroso.

With his book *Sur l'homme et le développement de ses facultés* (1835), Belgian astronomer, mathematician, statistician and sociologist Adolphe Quetelet became a major figure in statistical criminology. He linked biological and social normality to the frequency with which certain characteristics, including deviant ones, appeared in any given population. Quetelet used physics and astronomy, theology and religion to conduct statistical analyses of society's health and indirectly, crime. His analyses were based on the stability of the mean, rather than the dispersion of individual traits or events (Caponi, 2013). Meaning, deviance and crime of "the average man", according to Quetelet, would emerge from permanent causes that also give rise to law-abiding behaviour. Quetelet's propositions for a normal distribution of human traits laid the groundwork for Charles Darwin's argument on the operation of natural selection through the adequate variability of natural populations (Eiseley, 1958).

Influenced heavily by Darwin's *On the Origin of Species* (1859), Italian criminologist Cesare Lombroso wrote *L'uomo delinquente*, or *Criminal Man* (1876), which embraced the idea of evolution through natural selection. Lombroso considered criminals as a separate subspecies of humans, possessing distinct physical and mental characteristics such as uncommon skull sizes and unevenness of facial bones. He approached "criminals" as descendants of prehistoric man, or "biological throwbacks" subjected to what he conceptualised as atavism, or the reappearance of traits from an ancestor, for him a major factor in criminal behaviour.

This book is not going to entertain that humankind accommodates a "criminal subspecies" with distinct physical and mental characteristics. It actually rejects that "criminals" are a deviant biological variation of the human species, which is in and of itself a highly problematic assumption, as Rafter has shown (2008). Instead, here, the human species overall is considered from a criminologically imaginative point of view as a species with totally annihilative and exploitative tendencies, intentionally and unintentionally.

This is considered necessary because any criminology, especially a criminology of the human species, cannot begin without a criminological imagination, one that allows us to let go of conventional criminological epistemological and theoretical "beliefs" (Young, 2011). As the quote from Albert Einstein at the beginning of this section indicates, (crimino)logic takes us from A to Z, but the (criminological) imagination takes us everywhere. In fact, the imagination is considered to be one of the most powerful abilities humankind possesses (Hajer, 2017). Hence, without a criminological imagination of the human species, a criminology of the human species remains an unexplored topic, as state crime, corporate crime, genocide and ecocide once were (Cohen, 2013; Day & Vandiver, 2000; Friedrichs, 1994; South, 1998). Unlike those previously unexplored topics, the human species as a criminological topic is perhaps more difficult, taboo even, because it considers the possibility of the entire human species being deviant.

The criminological imagination helps ease the taboo. Being a criminological reproduction of Mills' sociological imagination (2000), the criminological imagination makes sense of dominant perceptions about crime and what they narrate about society. Crime and deviance are, as such, to be understood in relation to an understanding of socio-historical development of society itself, to deliver 'a set of viewpoints that are simple enough to make understanding possible, yet comprehensive enough to permit us to include in our views the range and depth of the human variety' (ibid.: 133).

The criminological imagination is therefore about understanding crime in a perspectival manner, enabled by a kaleidoscopic collection of a wide range of disciplinary perspectives, epistemologies, ontologies, theories and methodologies, addressing more parties than just the state (Bosworth & Hoyle, 2012). The criminological imagination ought to engage with a wide range of themes and sources, adhering to the interweaving of biography and history to inspire and challenge the field of criminology (Daems, 2006: 53). A criminological imagination of the human species is precisely about such delving into our species' biography and history, but also about possible futures. In that sense, a criminology of the human species challenges the field of criminology, because it must be freed from the criminological straitjackets that may exist, particularly when doing state-funded criminological research. A criminological imagination of the human species is one that is going to listen to the unheard that are hurt by the powerful. The powerful in this case, however, are not necessarily state or corporate actors (Barton et al., 2013: 208–209), but the entire human species itself. We are the most powerful species on Earth and we seem to be the only species on Earth that does not live in natural balance with its environment (anymore) (Eldredge, 2000; Rull, 2022; Walsh, 1984), whereas it is precisely that environment, filled with an unfathomable amount of flora and fauna, of other species that are the unheard. Also off-Earth.

Of course, all human activity cannot be disconnected from its specific material and historical contexts—meaning, the imagination here is not necessarily one of assigning all of the human species "blame". However, it does acknowledge that specific sectional interests of ruling classes could be linked to environmental and species destruction (Barnosky et al., 2011; Greshko, 2019; Leslie, 2002; Shearing, 2015; White, 2022), as much as there have been communal societies that have lived in relative harmony with nature over centuries (De Landa, 2000), as Chapter 4 of this book highlights. This book focuses on the deviant, fatal, and violent *tendencies* of the human species by which it is spiralling toward mass extinction while denying the impact on itself and other species. This denial is highly problematic, as argued by Lees et al. (2020) who advocate for scientists to fight the rise of mass extinction denial in public debates. A criminology of the human species is therefore necessary, working toward a criminological imagination to break with the denial of our exploitative and annihilating tendencies, paving the way for criminological research on the risk of accelerated mass extinction.

1.3   Against Criminology

To do so, the criminological imagination of the human species requires us to go against mainstream, conventional (and empiricist) criminology, which has a history of denying the most difficult topics, including state and/or corporate crime, war, genocide and environmental harms (Cohen, 2017). Critical criminology has addressed such silences and other types of deviance and harms that do not fit traditional perspectives on criminality and victimhood (Cohen, 2013; Day & Vandiver, 2000; Eski & Walklate, 2020; Friedrichs, 1994; South, 1998). The criminological imagination of the human species must go further. By spearheading an unconventional, unsettling and counterintuitive imagination, it builds on and enriches the critical criminological tradition, setting an unsettling tone for criminological research on the human species, our extinction and our destructive nature. This book is therefore an unsettling enrichment of existing, especially green, criminological approaches to deviance, crime and harm more broadly (Burke, 2020; Janssen & Schuilenburg, 2021; McClanahan, 2020; Natali, 2016; Van Uhm, 2018; White, 2021).

Although green criminology offers holistic analyses of crime and victimhood in relation to nature, in which 'all species are morally and ethically equal from a biocentric perspective' (Van Uhm, 2018: 42), this book pushes the imaginative boundaries further. Besides taking into account previous critique on dichotomous thinking that dominates criminology, such as the distinction between human/animal and perpetrator/victim (Janssen & Schuilenburg, 2021), this book enriches these critiques by prompting controversial questions about the human species. How does our possible extinction relate to previous mass extinctions? What is happening with humanity and being human in times of accelerated artificial intelligence (AI) developments, vis-à-vis the AI-fication of society? What about human enhancement innovations? To what extent is the human species a biospheric harm not only to other species, but also to the entire biosphere on this planet and potentially to extraterrestrial space and other possibly inhabitable planets? At what cost will our survival come in the future?

Wondering about these questions, the criminological imagination of the human species is a small but first and challenging step toward the development of an empirically informed criminology of the human species. It is a criminological imagination that interweaves biography and history, one in which we need to study the human species by drawing on various disciplines, such as archaeology, astronomy, astrophysics, biology, genetics, geology, physics, palaeontology, and planetology, as has been recommended in the study of mass extinctions (Pievani, 2014). These fields have examined the human species in ways that differ from criminology and their insights could enrich our understanding of our species' biography and history. Integration of these disciplines enables the delivery of a provocative, controversial imagination of the human species as a deviant, criminal and most destructive species (Glikson, 2021), one that in its interacting with nature has modified and keeps on modifying nature and itself, continually reshaping the terrain upon which and within which it emerges and exists (White, 2021: 76).

The criminological imagination to be delivered, then, is one that is freed from the stubborn belief that human life and the human condition are disconnected from nature and other species. The human species may very well not be eternal on or off Earth. Therefore, instead of fearing it, the criminological imagination of the human species embraces that:

> Without a world into which men are born and from which they die, there would be nothing but changeless eternal recurrence, the deathless ever-lastingness of the human as of all other animal species. (Arendt, 2019: 97)

Meaning, existing as a species is merely temporary. Our lives are transient—which we seem to have forgotten about—but we still act upon the idea of being an eternal species. What if we go extinct, or change into a being that is not human anymore? To answer these and similar key existential questions, and as will emerge from the topical-chronological Chapters 3–6, the criminological imagination of the human species is one in which 'neither the life of an individual [species] nor the [geological] history of [any biodiversity] can be understood without understanding both' (Mills, 2000: 3).

The criminological imagination of human species that is aimed for here thus dares criminologists to push their own criminological imagination of crime, harm and the human species beyond "the human being" itself, also by not shying away from post- and transhumanistic thinking (Bostrom, 2005; Drolet et al., 2020; Theiry et al., 2021). Readers are asked to challenge their assumptions and consider alternative perspectives that may be unsettling. If that is established, an unsettling criminological tone is set—one that then has made a criminological imagination of the human species a foundation toward more deviant "inconvenient" knowledge (Walters, 2003) on the human species.

1.4 Conclusion and Structure of the Book

As already stated, the criminological imagination of the human species will emerge from the subsequent chapters. The origin of the human species, which is one of mass exploitation and annihilation, will be the first element to be criminologically imagined (Chapter 3). The book then moves on to how the human species has become an all-encompassing, omnicidal violent force that has harmed Earth grossly and perhaps irreversibly by having self-inflicted the sixth (Anthropocene) mass extinction (Chapter 4). While we have enabled such mass extinction, the human species has simultaneously dismantled its own biology through human enhancement and working toward singularity: a hypothetical point in the future when the growth of (AI) technology becomes overpowering and irrevocable, leading to a dramatic and unpredictable transformation of our reality by means of intelligent and powerful technologies (Chapter 5). Finally, by expanding into space, the human species has gone to the final frontier of exploitation (Chapter 6). Prior to delving into these chapters, the book will explore the geological past and examine Earth's tumultuous pattern of mass extinctions, and the interconnected cycles of life, death, and rebirth within biodiversity.

REFERENCES

Arendt, H. (2019). *The human condition*. The University of Chicago Press.

Barnosky, A. D., et al. (2011). Has the earth's sixth mass extinction already arrived? *Nature, 471*(7336), 51–57.

Barton, A., Corteen, K., Scott, D., & Whyte, D. (2013). *Expanding the criminological imagination*. Routledge.

Bostrom, N. (2005). A history of transhumanist thought. *Journal of Evolution and Technology, 14*(1), 1–25.

Bosworth, M., & Hoyle, C. (2012). *What is criminology?* Oxford University Press.

Boutellier, H. (2019). *A criminology of moral order*. Bristol University Press.

Brisman, A., & South, N. (2013). A green-cultural criminology: An exploratory outline. *Crime Media Culture, 9*(2), 115–135.

Brisman, A., & South, N. (2020). A criminology of extinction: Biodiversity, extreme consumption and the vanity of species resurrection. *European Journal of Criminology, 17*(6), 918–935.

Broderick, M. (1993). Surviving armageddon: Beyond the imagination of disaster. *Science Fiction Studies, 20*(3), 362–382.

Burke, R. H. (2020). Green and species criminology. In R. H. Burke (Ed.), *Contemporary criminological theory* (pp. 300–323). Routledge.

Cairns, Jr., J. (2013). Can a species rapidly moving toward a self-inflicted extinction be considered successful? *Integrated Environmental Assessment and Management, 9*(4), 674–675.

Campbell, D. T. (1974). 'Downward causation' in hierarchically organised biological systems. In *Studies in the philosophy of biology* (pp. 179–186). Palgrave.

Caponi, S. (2013). Quetelet, the average man and medical knowledge. *História, Ciências, Saúde-Manguinhos, 20*, 830–847.

Chowdhury, R. B., et al. (2021). Environmental externalities of the COVID-19 lockdown: Insights for sustainability planning in the Anthropocene. *Science of the Total Environment, 783*, 147015.

Clark, M. E. (1989). Humankind at the crossroads. In *Ariadne's thread* (pp. 471–506). Palgrave Macmillan.

Cohen, S. (2013). *States of denial: Knowing about atrocities and suffering*. Wiley.

Cohen, S. (2017). *Against criminology*. Routledge.

Colossal. (2023). *The mammoth*. https://colossal.com/mammoth/

Daems, T. (2006). Zygmunt Bauman en de criminologische verbeelding. *Panopticon: tijdschrift voor strafrecht, criminologie en forensisch welzijnswerk, 27*(6), 51–54.

Darwin, C. (1859). *The origin of species by means of natural selection, or the preservation of favoured races in the struggle for life*. John Murray.

Day, L. E., & Vandiver, M. (2000). Criminology and genocide studies: Notes on what might have been and what still could be. *Crime, Law and Social Change, 34*(1), 43–59.

De Landa, M. (2000). *A thousand years of nonlinear history*. Swerve Editions.

Doyle, B. (2015). The postapocalyptic imagination. *Thesis Eleven, 131*(1), 99–113.

Drolet, M. J., et al. (2020). Intergenerational occupational justice: Ethically reflecting on climate crisis. *Journal of Occupational Science, 27*(3), 417–431.

Dunnett, O., Maclaren, A., Klinger, J., Maria, K., Lane, D., & Sage, D. (2019). Geographies of outer space: Progress and new opportunities. *Progress in Human Geography, 43*(2), 314–336.

Eiseley, L. (1958). *Darwin's century*. Doubleday.

Eldredge, N. (2000). *Life in the balance: Humanity and the biodiversity crisis*. Princeton University Press.

Eski, Y. (2022). Omnia cadunt. Naar een victimologische verbeelding van onze vergankelijkheid Omnia cadunt [Toward a victimological imagination of our transience]. *Tijdschrift over Cultuur & Criminaliteit, 12*(1), 58–71.

Eski, Y., & Walklate, S. (2020). A victimological imagination of genocide. In Y. Eski (Ed.), *Genocide and victimology* (pp. 202–210). Routledge.

Ferrell, J., Hayward, K., Morrison, W., & Presdee, M. (2004). *Cultural criminology unleashed*. Routledge.

Ferrell, J., & Sanders, C. (1995). *Cultural criminology: An invitation*. UPNE.

Friedrichs, D. O. (1994). Crime wars and peacemaking criminology. *Peace Review, 6*(2), 159–164.

Galey, P. (2022). *2022 was Europe's hottest summer on record by a 'substantial margin'*. https://www.sciencealert.com/2022-was-europes-hottest-summer-on-record-by-a-substantial-margin

Garland, D. (1992). Criminological knowledge and its relation to power: Foucault's genealogy and criminology today. *British Journal of Criminology, 32*, 403–422.

Gee, H. (2021). *Humans are doomed to go extinct.* https://www.scientificameri can.com/article/humans-are-doomed-to-go-extinct/

Glikson, A. Y. (2021). *The fatal species: From warlike primates to planetary mass extinction.* Springer.

Green, P. J., & Ward, T. (2000). State crime, human rights, and the limits of criminology. *Social Justice, 27*(1), 101–115.

Greshko, M. (2019). *Wat waren de 5 massa-uitstervings en wat veroorzaakte ze? Van.* https://www.nationalgeographic.nl/wetenschap/2019/09/wat-waren-de-5-massa-uitstervings-en-wat-veroorzaakte-ze

Hajer, M. (2017). *The power of imagination.* Universiteit Utrecht.

Hayward, K. (2016). Space–the final frontier: Criminology, the city and the spatial dynamics of exclusion. In *Cultural criminology unleashed* (pp. 169–180). Routledge.

Hirschfeld, A. R., & Blackmer, S. (2021). Beyond acedia and wrath: Life during the climate apocalypse. *Anglican Theological Review, 103*(2), 196–207.

IPCC. (2021). *Climate change 2021: The physical science basis report.* https://www.ipcc.ch/report/ar6/wg1/downloads/report/IPCC_AR6_WGI_SPM.pdf

Janssen, J., & Schuilenburg, M. (2021). Het antropoceen. De criminologische uitdaging in de 21ste eeuw. *Tijdschrift over Cultuur & Criminaliteit, 11*(1), 3–13.

Lampkin, J. (2021). *Should criminologists be concerned with outer space? A proposal for an 'astro-criminology'.* https://research.leedstrinity.ac.uk/en/pub lications/should-criminologists-be-concerned-with-outer-space-a-proposal-fo

Lasch, C. (1979). *The culture of narcissism.* Norton.

Lees, A. C., Attwood, S., Barlow, J., & Phalan, B. (2020). Biodiversity scientists must fight the creeping rise of extinction denial. *Nature Ecology & Evolution, 4*(11), 1440–1443.

Leslie, J. (2002). *The end of the world: The science and ethics of human extinction.* Routledge.

Lombroso, C. (1876). *L'uomo delinquent. Studiato in rapporto alla antropologia, alla medicina legale ed alle discipline carcerarie.* Ulrico Hoepli, Libraio-editore.

Maynard, A. (2022). *'Jurassic World' scientists still haven't learned that just because you can doesn't mean you should—Real-world genetic engineers can learn from the cautionary tale.* https://theconversation.com/jurassic-world-scientists-still-havent-learned-that-just-because-you-can-doesnt-mean-you-sho uld-real-world-genetic-engineers-can-learn-from-the-cautionary-tale-184369

McClanahan, B. (2020). Earth–world–planet: Rural ecologies of horror and dark green criminology. *Theoretical Criminology, 24*(4), 633–650.

McGarry, R., & Walklate, S. (2019). *A criminology of war?* Bristol University Press.

Mills, C. W. (2000). *The sociological imagination.* Oxford University Press.

NASA. (2022). *First images from the James Webb space telescope.* https://www.nasa.gov/webbfirstimages.

NASA. (2023). *NASA to launch new mars sample receiving project office at Johnson.* https://www.nasa.gov/press-release/nasa-to-launch-new-mars-sample-receiving-project-office-at-johnson

Natali, L. (2016). Green criminology with eyes wide open. In *A visual approach for green criminology* (pp. 1–14). Palgrave.

Norgaard, K. M. (2020). Whose energy future? Whose imagination? Revitalizing sociological theory in the service of human survival. *Society & Natural Resources, 33*(11), 1438–1445.

Padilla, L. A. (2021). The Anthropocene: Are We in the midst of the sixth mass extinction? In *Sustainable development in the Anthropocene* (pp. 93–167). Springer.

Pease, K. (2021). Commentary to "how international are the top ten international journals of criminology and criminal justice?" *European Journal on Criminal Policy and Research, 27*, 179–182.

Persinger, K. (2020). Constructing reality: An investigation of climate change and the terraforming imaginary. *The Macksey Journal, 1*(1), 1–16.

Pievani, T. (2014). The sixth mass extinction: Anthropocene and the human impact on biodiversity. *Rendiconti Lincei, 25*(1), 85–93.

Quetelet, A. (1835). *Sur l'homme et le développement de ses facultés. Essai de Physique Sociale.* Bachelier, Imprimeur-Libraire.

Rafter, N. (2008). Criminology's darkest hour: Biocriminology in Nazi Germany. *Australian & New Zealand Journal of Criminology, 41*(2), 287–306.

Revive & Restore. (2023). *Woolly mammoth revival.* https://reviverestore.org/projects/woolly-mammoth/

Rull, V. (2022). Biodiversity crisis or sixth mass extinction? Does the current anthropogenic biodiversity crisis really qualify as a mass extinction? *EMBO Reports, 23*(1), e54193.

Shearing, C. (2015). Criminology and the Anthropocene. *Criminology & Criminal Justice, 15*(3), 255–269.

Slapper, G., & Tombs, S. (1999). *Corporate crime.* Longman.

South, N. (1998). A green field for criminology? A proposal for a perspective. *Theoretical Criminology, 2*(2), 211–233.

Szocik, K. (2019). *The human factor in a mission to mars: An interdisciplinary approach.* Springer.

Theiry, W., et al. (2021). Intergenerational inequities in exposure to climate extremes. *Science.* https://doi.org/10.1126/science.abi7339

Tombs, S. (2018). For pragmatism and politics: Crime, social harm and zemiology. In A. Boukli & J. Kotzé (Eds.), *Zemiology* (pp. 11–31). Palgrave Macmillan.

Trevorrow, C. (2022). *Jurassic World: Dominion.* Amblin Entertainment et al./Universal Pictures.

Van Uhm, D. P. (2018). Naar een non-antropocentrische criminologie. *Tijdschrift over Cultuur En Criminaliteit, 1,* 35–53.

Walsh, R. (1984). *Staying alive: The psychology of human survival.* Shambhala Publications.

Walters, R. (2003). *Deviant knowledge.* Willan.

White, R. (2021). *Theorising green criminology: Selected essays.* Routledge.

White, R. (2022). Climate change and the geographies of ecocide. In M. Bowden & A. Harkness (Eds.), *Rural transformations and rural crime* (pp. 108–124). Bristol University Press.

Young, J. (2011). *The criminological imagination.* Polity Press.

CHAPTER 2

Circles of Life and Death: Previous Biodiversity and Mass Extinctions

Abstract This chapter looks back beyond our origins to consider the origin of Earth and its evolution comprising five completed mass extinctions and (re)births of previous biodiversity. Earth's origin and its long history has been a coming and going of entire other biospheres and the species they accommodated, rendering the current sixth mass extinction both real and unexceptional. A criminological imagination of that geological biography and history delivers a contextual understanding of how extinction is not deviant but normal, and is in line with Earth's circles of life and death.

Keywords Mass extinctions · Biodiversity · Cosmic violence

2.1   A Brutal Cosmic Origin

You are a child of the universe, no less than the trees and the stars. (Max Ehrmann, 1948: 165)

It has been a well-established and acknowledged theory that, as far as we know, the universe was shaped by a big bang (Lemaître, 1931 [2013]). It

© The Author(s), under exclusive license to Springer Nature Switzerland AG 2023
Y. Eski, *A Criminology of the Human Species*,
Palgrave Studies in Green Criminology,
https://doi.org/10.1007/978-3-031-36092-3_2

was an explosion of energetic violence from which a universal structure of superclusters, clusters, galaxy groups, and subsequently, galaxies formed themselves, like the Milky Way in which our solar system lies. Out of that galactic violence came about

> …the chemical elements of life [that] were first produced in the first gener-ation of stars after the Big Bang. […] One way or another the first stars must have influenced our own history, beginning with stirring up every-thing and producing the other chemical elements besides hydrogen and helium. So if we really want to know where our atoms came from, and how the little planet Earth came to be capable of supporting life, we need to measure what happened at the beginning. (Mather in NASA, 2022—online source)

The Milky Way resulted from a dramatic history of violence, having had about 13.6 billion years of galactic mergers and collisions, and even complete consumption of multiple other galaxies (Malhan et al., 2022). The solar system first saw the light approximately 4.6 billion years ago (Bouvier & Wadhwa, 2010). Not "shortly" after that, Earth formed around 4.5 billion years ago in that early solar system, along with the rest of the planets, moons, asteroids and comets. The Sun formed in the centre of a nebula as gravity collapsed the material and caused it to spin, condensing the matter (Cameron, 1985). This process also created gas giants and terrestrial planets, including Earth.

As Earth's rocky core moulded itself, its magnetic field may have formed around this time as well (Kargel & Lewis, 1993). It has been proposed that an impact on Earth caused the formation of the Moon from Earth's mantle, which has had a significant influence on Earth's evolution (Barboni et al., 2017). Plate tectonics and volcanism shaped Earth's surface and played a role in the development of the atmosphere (Hawkesworth et al., 2010). Water on Earth's surface may have come from collisions with comets and asteroids, which made our planet a region where liquid water, a key ingredient for life, could exist (Albertsson et al., 2014; Brack, 1993). Some of these key life ingredients on Earth have been found in meteorites coming from space (Oba et al., 2022).

This means that our bodies do really mirror the universe, so 'when we look at very distant galaxies, we see the same components that we are made of carbon, hydrogen, iron', as University of Cambridge astronomer Matthew Bothwell explained (Khan, 2022). All life on Earth consists of matter derived from celestial bodies exploding and colliding. Therefore, when it is claimed we are all made of stardust, it literally means that human life too stems from the cosmic carnage of aggressive planetary collisions, hostile mergers and especially massive star explosions, called supernovae (Suresh & Kumar, 2005).

Exploding and imploding stars have ignited life, as much as they have also caused mass extinctions through climate change, resulting in the loss of megafauna, such as the megalodon (Melott et al., 2019), as well as influencing human evolution on an existential level (Svensmark, 2022). Therefore, we are not just made up of the same matter as Earth, other planets, comets, asteroids, and the universe (Kerns, 2017), but life on Earth has also been shaped by cosmic violence, a process of life and death that has cyclically continued throughout the planet's history.

2.2 Earth's Life of Mass Extinctions

When inspecting Earth's surface, and the crusts and layers further down, we discover the geological history of Earth. The layered stack of Doonbristy Rock in Ireland (see Illustration 2.1), for example, makes geological history visible (Graham, 1996). These layers tell the story of Earth's biography encoded in traces (or scars) of previous, mass-extinct life.

Throughout geological history, multiple biodiversities on our planet have existed and have become extinct during five mass extinctions. We now reside in the sixth mass extinction (Greshko, 2019). About five hundred million years ago, first life began and then ended, after which new life was reborn (cf. Luo et al., 2020; Pievani, 2014). Mass extinctions ensure such cycles of life and death, and are thus part of a larger circular process that has been repeated throughout geological history:

1. the Late Ordovician mass extinction, 447–444 million years ago;
2. the Late Devonian extinction, 374–359 million years ago;
3. the Permian–Triassic mass extinction event, also called the Great Extinction, 251 million years ago;
4. the Triassic–Jurassic extinction event, 202 million years ago, and;
5. the Cretaceous–Tertiary boundary, 65–95 million years ago, when almost all dinosaurs became extinct.

Illustration 2.1 Doonbristy Rock (*Source* https://upload.wikimedia.org/wik
ipedia/commons/2/25/Dun_Briste%2C_Downpatrick_Head_-_geograph.org.
uk_-_369248.jpg?20110203115043)

The legacy of these mass extinctions can literally be found in Earth's crust,
in nickel isotopes, for example (Li et al., 2021), as well as in plant and
animal fossils (Luo et al., 2020). This geo-historical registration in the
Earth's crust shows that each of these mass extinctions had its own characteristics, but also overlapping ones. For example, when the first mass
extinction happened, there were drastic changes due to massive cooling.
Countless species became extinct because they could not survive in inhos-
pitable waters, making the first mass extinction the second most severe;
about 85% of all species were wiped out, especially marine organisms
(Stigall, 2019). During the second mass extinction, around 360 million
years ago, extraterrestrial impacts by asteroids and comets, and large-scale
volcanism, mainly targeted marine animals (Percival et al., 2018). It shows
that mass extinctions have both terrestrial and extraterrestrial causes, as is
written in our Earth's crust.

The most destructive mass extinction was the third one, the Permian–
Triassic. An estimated 96% of all marine species and about 75% of
all terrestrial species went extinct within just 60,000 years. Forests
completely disappeared and only returned ten million years later, just as

some insect species, such as marine ecosystems, took almost eight million years to recover (Stigall, 2019). One of the main causes of this Great Extinction, as it is called, were the volcanic eruptive fissures that covered present-day Siberia with lava, emitting at least 14,500 billion tons of CO_2 (Payne & Clapham, 2012), resulting in a planet warmed to a degree that made survival unimaginable. If we were to compare that to today, that amount of greenhouse gases is more than twice as large as the amount that would be released if we completely used up all the planet's oil and gas (Greshko, 2019). While we are now concerned that we do not want global warming to go above 1.5 degrees, during the third great mass extinction the temperature rose by 14–20 degrees. The sea would have felt like a hot bath and it rained sulphur ash (Benton, 2018). This clarifies that carbon (dust) dioxide emissions are the main cause of global warming, reflected in the acidification of waters, as a result of which oxygen (exponentially) decreased rapidly. It also reveals that Earth has its own geological logic of life and death, in which volcanoes not only terminate life, but also provide phosphorus that enables life through microbes that make oxygen (Meixnerová et al., 2021). Geological change through volcanism takes and gives life, as is said of Mount Etna in Sicily (TripAdvisor, 2016).

The fourth mass extinction, which was the Triassic–Jurassic extinction event, comprised a mass extinction that took place approximately 202 million years ago and was completed in a mere 10,000 years. It had a significant impact on the development of life on land and in particular in the sea, where roughly 20% of all marine families passed away (Kiessling & Aberhan, 2007). As with all mass extinctions, several hypotheses exist about how the Triassic–Jurassic extinction could have happened,: gradual climate change, meteorite impact, volcanism or increased methane in the atmosphere, leading to a rapid global warming (Hesselbo et al., 2002). None of these theories are entirely conclusive, and it remains unclear which one is strongest.

The fifth and so far final mass extinction wiped out 76% of all life on Earth, including nearly all dinosaurs. It was previously claimed that they became extinct due to their own slowness and clumsiness, but that could not explain the fairly rapid transition of these reptiles to mammals (Alvarez et al., 1980). In the 1970s and 1980s there was little room in the imagination for the idea that dinosaurs became extinct due to an asteroid impact about 66 million years ago off the coast of Mexico (Greshko, 2019). This idea was received with much scepticism and even ridiculed by palaeontologists and geologists (Pievani, 2014), but was later accepted. It is now accepted that the dinosaurs became extinct because of the asteroid, but also because of declining biodiversity due to the cooling of the Earth (Condamine et al., 2021).

2.3   Conclusion

In this chapter, we looked back beyond our origins to the most violent beginnings of Earth and life on Earth, and then discussed the previous five mass extinctions and mass (re)births of biodiversity and its evolution. The main lesson to be drawn from this chapter is that Earth's origin and its history comprise a coming and going of entire, previous biospheres and the species that lived back then. Out of violence comes life, which is violently taken away, in order to bring new life. It makes the current sixth mass extinction, the Anthropocene mass extinction, far from unique. What is unique though, as will be explored in more detail in Chapter 4, is that the human species has caused and accelerated the Anthropocene mass extinction. It makes us a most violently destructive species, ever since we came into existence, as will be elaborated on in the next chapter.

References

Albertsson, T., Semenov, D., & Henning, T. (2014). Chemodynamical deuterium fractionation in the early solar nebula: The origin of water on earth and in asteroids and comets. *The Astrophysical Journal, 784*(1), 39.

Alvarez, L. W., et al. (1980). Extraterrestrial cause for the Cretaceous-Tertiary extinction. *Science, 208*(4448), 1095–1108.

Barboni, M., Boehnke, P., Keller, B., Kohl, I. E., Schoene, B., Young, E. D., & McKeegan, K. D. (2017). Early formation of the moon 4.51 billion years ago. *Science Advances, 3*(1), e1602365.

Benton, M. J. (2018). Hyperthermal-driven mass extinctions: Killing models during the Permian-Triassic mass extinction. *Philosophical Transactions of the Royal Society A: Mathematical, Physical and Engineering Sciences, 376*(2130), 20170076.

Bouvier, A., & Wadhwa, M. (2010). The age of the solar system redefined by the oldest Pb–Pb age of a meteoritic inclusion. *Nature Geoscience, 3*(9), 637–641.

Brack, A. (1993). Liquid water and the origin of life. *Origins of Life and Evolution of the Biosphere, 23*(1), 3–10.

Cameron, A. G. W. (1985). Formation and evolution of the primitive solar nebula. *Protostars and Planets II*, 1073.

Condamine, F. L., et al. (2021). Dinosaur biodiversity declined well before the asteroid impact, influenced by ecological and environmental pressures. *Nature Communications, 12*(1), 1–16.

Ehrmann, M. (1948). *The desiderata of love: A collection of poems for the beloved.* Crown.

Graham, J. R. (1996). Dinantian river systems and coastal zone sedimentation in northwest Ireland. *Geological Society, London, Special Publications, 107*(1), 183–206.

Greshko, M. (2019). *Wat waren de 5 massa-uitstervings en wat veroorzaakte ze? Van.* https://www.nationalgeographic.nl/wetenschap/2019/09/wat-waren-de-5-massa-uitstervings-en-wat-veroorzaakte-ze

Hawkesworth, C. J., Dhuime, B., Pietranik, A. B., Cawood, P. A., Kemp, A. I., & Storey, C. D. (2010). The generation and evolution of the continental crust. *Journal of the Geological Society, 167*(2), 229–248.

Hesselbo, S. P., Robinson, S. A., Surlyk, F., & Piasecki, S. (2002). Terrestrial and marine extinction at the Triassic-Jurassic boundary synchronized with major carbon-cycle perturbation: A link to initiation of massive volcanism? *Geology, 30*(3), 251–254.

Kargel, J. S., & Lewis, J. S. (1993). The composition and early evolution of earth. *Icarus, 105*(1), 1–25.

Kerns, J. (2017). Mining materials in outer space. *Machine Design, 89*(5), 9–10.

Khan, C. (2022). *Will we ever see pictures of the big bang? We ask an expert.* https://www.theguardian.com/lifeandstyle/2022/sep/23/will-we-ever-see-pictures-of-the-big-bang-we-ask-an-expert

Kiessling, W., & Aberhan, M. (2007). Geographical distribution and extinction risk: Lessons from Triassic-Jurassic marine benthic organisms. *Journal of Biogeography, 34*(9), 1473–1489.

Lemaître, G. (1931 [2013]). A homogeneous universe of constant mass and increasing radius accounting for the radial velocity of extra-galactic nebulae. In *A source book in astronomy and astrophysics, 1900–1975.* Harvard University Press.

Li, M., Grasby, S. E., Wang, S. J., Zhang, X., Wasylenki, L. E., Xu, Y., ... & Shen, Y. (2021). Nickel isotopes link Siberian Traps aerosol particles to the end-Permian mass extinction. *Nature Communications, 12*(1), 2024.

Luo, M., et al. (2020). Trace fossils as proxy for biotic recovery after the end-Permian mass extinction: A critical review. *Earth-Science Reviews, 203*, 103059.

Malhan, K., Ibata, R. A., Sharma, S., Famaey, B., Bellazzini, M., Carlberg, R. G., D'Souza, R., Yuan, Z., Martin, N. F., & Thomas, G. F. (2022). The global dynamical atlas of the milky way mergers: Constraints from Gaia EDR3–based orbits of globular clusters, stellar streams, and satellite galaxies. *The Astrophysical Journal, 926*(2), 107.

Meixnerová, J., et al. (2021). Mercury abundance and isotopic composition indicate subaerial volcanism prior to the end-Archean "whiff" of oxygen. *Proceedings of the National Academy of Sciences, 118*(33), e2107511118.

Melott, A. L., Marinho, F., & Paulucci, L. (2019). Hypothesis: Muon radiation dose and marine megafaunal extinction at the end-Pliocene supernova. *Astrobiology, 19*(6), 825–830.

NASA. (2022). *Webb telescope and the Big Bang.* https://webb.nasa.gov/content/features/bigBangQandA.html

Oba, Y., Takano, Y., Furukawa, Y., Koga, T., Glavin, D. P., Dworkin, J. P., & Naraoka, H. (2022). Identifying the wide diversity of extraterrestrial purine and pyrimidine nucleobases in carbonaceous meteorites. *Nature Communications, 13*(1), 2008.

Payne, J. L., & Clapham, M. E. (2012). End-Permian mass extinction in the oceans: An ancient analog for the twenty-first century?. *Annual Review of Earth and Planetary Sciences, 40*, 89–111.

Percival, L. M., Jenkyns, H. C., Mather, T. A., Dickson, A. J., Batenburg, S. J., Ruhl, M., Hesselbo, S. P., Barclay, R., Jarvis, I., Robinson, S. A., & Woelders, L. (2018). Does large igneous province volcanism always perturb the mercury cycle? Comparing the records of Oceanic Anoxic Event 2 and the end-cretaceous to other Mesozoic events. *American Journal of Science, 318*(8), 799–860.

Pievani, T. (2014). The sixth mass extinction: Anthropocene and the human impact on biodiversity. *Rendiconti Lincei, 25*(1), 85–93.

Stigall, A. L. (2019). The invasion hierarchy: Ecological and evolutionary consequences of invasions in the fossil record. *Annual Review of Ecology, Evolution, and Systematics, 50*, 355–380.

Suresh, P. K., & Kumar, V. H. (2005). Supernovae: Explosions in the Cosmos. ArXiv: astro-ph/0504597.

Svensmark, H. (2022). Supernova rates and burial of organic matter. *Geophysical Research Letters, 49*(1), e2021GL096376.

TripAdvisor. (2016). *The Etna gives, the Etna takes....* https://www.tripadvisor.com/ShowUserReviews-g660768-d10666654-r408182618-EtnaToursfromTaormina-Linguaglossa_Province_of_Catania_Sicily.html

CHAPTER 3

Born of Violence: The Neanderthal Extinction, Genocide and Colonisation

Abstract In this chapter, we imagine how the unique human capacities of the opposable thumb and the precision of our brain have enabled us to imagine and (physically) bring about progress. However, these same capacities have also allowed us to exploit and annihilate other human beings and human-like species on a mass scale. Specifically, the Neanderthal. At the core of this ability is our capacity to imaginatively dehumanise others and use this dehumanisation to justify and eventually materialise genocide and colonisation.

Keywords Neanderthal · Genocide · Colonisation · Dehumanisation

3.1 Mirror, Mirror on the Wall, Who's the Most Violent Animal of Them All?

Although mosquitos are the deadliest species to human beings, as is often suggested (cf. Douglas, 2022; Gates, 2014; Williams, 2022), perhaps one of the most dangerous and deadliest species to the entire planet is the human species. Not only do we deviate because other species live

© The Author(s), under exclusive license to Springer Nature Switzerland AG 2023
Y. Eski, *A Criminology of the Human Species*,
Palgrave Studies in Green Criminology,
https://doi.org/10.1007/978-3-031-36092-3_3

in balance with their ecology and we do not; also anatomically, physio-logically, and cognitively, the human species deviates from other species (Weck, 2022).

Compared to most species, our violence is different from animal violence, first of all, because we can grab more and more precisely than other species. The opposable thumb gives us the ability to engage in pad-to-pad grasping, which is a significant feature of our hands that distinguishes us from other primates (id.). Other apes possess oppos-able thumbs too, but those are optimised for tree-climbing, whereas for human beings the thumb functions for precision manipulation of objects (Napier, 1956; Rolian, 2016); we have wider ranges of thumb move-ment. Even if the normal ranges of movements for the human thumb joints are reduced, that reduction in thumb joint movement is compen-sated for by an increased movement range in the other joints (Barakat et al., 2013). This means that whether we like it or not, our opposable thumb is designed for precision work.

More importantly, the same goes for our brains. Not because we have bigger brains than most species (e.g. elephants and whales have bigger ones), but because we have a more delicate brain that gives us mental precision. Our brain consists of far more neurons and networked inter-connections between them than other species' brains. We possess higher cognitive thinking, like reasoning logically and abstract thinking (Azevedo et al., 2009). The neuronal difference between humans and other species is caused by us having a neuron with far fewer ion channels; it has devel-oped the human brain into being able to divert energy to other neural processes (Beaulieu-Laroche et al., 2021). Other, in particular mammalian species, have a different building plan, in which it seems that the cortex is trying to keep the numbers of ion channels per unit volume the same across all the species. That would mean that for a given volume of cortex, the energetic cost is and remains the same, at least for ion channels (id.). We do not have an increased density of ion channels, but rather a dramat-ically decreased density of ion channels for a given volume of brain tissue. It implies that human brains have evolved in such a way that we require less energy for pumping ions and more energy for something else (id). Such energy is then redirected to creating more complicated synaptic connections between neurons, for example. It suggests that as a species, we evolved out of a building plan that restricted the size of cortex and evolved into having a brain that has become more energetically efficient compared to other species. Our brain and neuronal evolution have led to a hardwiring of precision and efficiency thinking, literally. This means that the human brain deviates from the building plan of other mammalian species, which cannot achieve the same level of precision and efficiency in their thinking processes.

In summary, we are able to grab objects more precisely than other species due to our opposable, precision-manipulative thumbs. This precision manipulation, together with our increased precision mental abilities, is what makes us deviant in comparison to other species and has contributed to our status as supposedly the most evolved and advanced species on the planet (Kumar, 2012; Troxell, 1936). However, as Solís (2004: 341) points out, 'while human beings are the most intelligent living species, they are also the most predatory'. The human species is characterised by its capacity for violence and destruction, and we are weaponised with 'hands, nails, and teeth, as well as stones, pieces of wood, flames, and fire, as soon as they are known' (Lucretius in Eski, 2022b: 3). Archaeologists have found stone tips of arrows and spears used for killing over 60,000 years ago (Lombard & Phillipson, 2010), after which weapons were eventually adapted for farming purposes, such as ploughing fertile soil (cf. Wells, 2011). These precision tools, whether for killing or ploughing, were made by human hands and functioned as extensions of the physical weapons attached to the human body (Young, 2003). We have mastered the weaponisation of the materials around us to destroy and exploit other human beings as well as other beings similar to us. Our precision and efficiency-oriented hands and brains are uniquely human aspects that, in comparison to other species, make us highly effective in our tendencies toward total exploitation and annihilation, right from the beginning.

3.2 Our Species' Criminogenesis: The Neanderthal Genocide

The crime of crimes, genocide, and its connection to colonialism, is intrinsically connected to the origins of humanity; genocidal violence and mass exploitation are perhaps *the* defining characteristics of being human. Still, whenever we imagine genocide, we think of the Holocaust, when Hitler's Nazi regime structurally annihilated 11 million Jews, Sinti-Roma peoples, homosexuals, political enemies and many other groups considered "inferior" by the Nazi regime. We see the Holocaust as exceptional, performed by exceptional human beings, or rather, by demons, monsters, who were inhuman. However, the Holocaust was enabled by very normal citizens. Arendt's controversial report (2006) on Nazi SS Leader Adolf Eichmann, the official who logistically organised the train transports to extermination and concentration camps, revealed that Eichmann's behaviour was normal. He did not have any evil instincts; it was rather bureaucracy and obedience—the "*Befehl ist Befehl*" ("An order is an order") mentality—that made him do his everyday work, very efficiently (id. 135). Although acknowledged and accepted in the meantime (cf. Browning, 1992; Katz, 1993), when Arendt's analysis of Eichmann as an ordinary man was published, it 'outraged people because it removed from monstrous crimes the depth and darkness to which we usually consign them' (Donoghue, 1979: 283–284). The Milgram experiment (1963) on authority and obedience, arguably, supports Arendt's thesis that evil is normal and banal, and thus, a will to total annihilation is normal.

Here, the criminological imagination of the human species pushes Arendt's banality of evil thesis ever further. Total annihilation is not only normal under certain circumstances; it is at the core of our being, at the very origin of the human species. The beginning of humankind is a totally annihilative criminogenesis. Next to leaving behind regular patterns of rapid extinction of animals throughout history wherever we set foot (Smith et al., 2018), one of the first extinctions caused by the human species has been that of a human-like being: the Neanderthal (Best, 2009; Caygill, 2016; Diamond, 2013; Kochi & Ordan, 2008; Walsh, 2010). However, before getting to how the Neanderthal exodus came about due to human activity, first a short shared history between Neanderthals and human beings.

Initial Upper Palaeolithic assemblages, including shell beads and bone artefacts, may have been produced by modern humans before they left Africa and spread to Europe as early as 50,000 years ago. This early arrival in western Eurasia suggests patchy colonisation with overlap in time between modern humans and Neanderthals, and the possibility of cultural diffusion from modern humans into the Neanderthal world – in other words, modern humans were unable to completely replace Neanderthals in all regions where they lived (Hublin, 2012). We actually shared many features with the Neanderthals, like rituals, hunting gear, and hunting the same prey and fighting over it, and we have even interbred with Neanderthals, as on average the human species possesses 2% of Neanderthal DNA (Galway-Witham et al., 2019).

Recently, it has been discovered that in the area of eastern central Germany, and about 125,000 years ago, Neanderthals gathered to butcher massive elephants; such large-scale elephant hunting required a degree of organisation(Gaudzinski-Windheuser et al., 2023). A lot of elephant meat was gathered this way, feeding 100 Neanderthal people for a month (id.), which implies large seasonal gatherings took place and food was stored (Curry, 2023—online source) and that indicates that

> ...these human ancestors were more sophisticated than once assumed, capable of adapting their behavior to a wide variety of environments and climates. [...] This lets us imagine Neanderthals as more like modern humans rather than as humanoid brutes, as they once were interpreted. (id.)

Although humankind and Neanderthals looked and lived alike, it is the human species that may have caused and/or accelerated the physical extinction of the Neanderthal (Caygill, 2016; Walsh, 2010). There are several theories about the human species stimulating the Neanderthal extinction. First of all, with the spread of diseases in Europe from Africa, the Neanderthal population was infected and had virtually no defence against them (Greenbaum et al., 2019), thus favouring human beings. These infectious diseases that affected Neanderthals also gave modern humans an advantage in terms of overcoming the spread of diseases into Eurasia (id.). It has also been argued that human beings came to dominate Neanderthals due to our use of archery, whereas Neanderthals kept on using stone-tipped speers (Metz et al., 2023). Bow and arrow enabled human beings to stay at a safe distance from large game, like bison, and allowed for precision kills; spears do not have the same advantages (id.). Given that the human species monopolised the catching of food, the Neanderthals were killed by mass starvation, resembling the Holodomor, a man-made famine in Soviet Ukraine that killed millions of Ukrainians in 1932–33 (Bezo & Maggi, 2015). Whether intentionally or not, the human species seems to have assisted in starving out another humanoid species.

These analytic speculations make interesting explanatory angles to criminologically imagine the human species' origin and its humanity, which is exactly what philosopher George Bataille did (Caygill, 2016). In his imaginations about what humanity is, he often struggled with the question whether (a) humans drove Neanderthal to extinction by inter-breeding, from which a new humanity came forward, or (b) the homo sapiens from Africa invaded Neanderthal territory—present-day Europe—and massacred the Neanderthal species. Bataille eventually opted for the Neanderthal genocide possibility (id. 264). It 'raises questions about the distinction between an act of nature, in the context of the struggle for survival, and an act of culture, in this case, genocide' (Walsh, 2010: 2). Imaginably, our totally annihilative origin can thus be considered cultur-ally embedded in our species, perhaps as 'a proto-cultural evolutionary struggle for survival' (id.). Either way, one of the first inflicted extinctions by humankind is that of another sapient, humanoid being.

Perhaps the Neanderthal extinction should be acknowledged as a genocide, strengthening the idea that 'the human heritage—and the prop-agation of itself as a thing of value—has occurred on the back of seemingly endless acts of violence, destruction, killing and genocide' (Kochi & Ordan, 2008: 11). It would mean that our species' genocidal intent is not just something normal under certain circumstances, as Arendt (2006) would have it, but that it is a defining element of the human species (Best, 2009; Diamond, 2013). The human species, then, possesses at least one 'consistent pattern reaching back into our early history and show[ing] a systematic problem inherent in our species itself—a proclivity toward violence that is likely to abide whatever the social setting' (Best, 2009: 291). Meaning that 'of all our human hallmarks—art, spoken language, drugs, and the others—the one that has been derived most straightforwardly from animal precursors is genocide' (Diamond, 2013: 264).

Imagining that the human species arose from the ashes of another sapient species, the Neanderthals (Banks et al., 2008; Eski, 2020; Golding, 1955; Jones, 2016) makes our birthright one of human-made extinction of another humanoid. We are 'the invasive species and agent of mass extinction par excellence' (Best, 2009: 298) and we are capable of dissolving other sapient (humanoid) species in our attempts to survive and thrive ourselves (Jones, 2016), and indeed, to *exist*. The annihilation of the Neanderthals has a wide range of ethical implications; however, it is an almost impossible task…

> …to locate this event on a scale of evils. It is more than genocide, because what was lost was not a race of human beings, but an entire sapient species. It is also more than the extinction of a species, because it was a species with

(presumably) a 'humanity' and self-consciousness somewhat like our own. (Silas SSF, 1997: 75)

Although more than 45,000 years ago, '[t]he first act of human genocide, [was] the massacre of the Neanderthals, [...] a prelude to the assault of agricultural societies against primal peoples for thousands of years' (Best, 2009: 291), and due to potential interbreeding between homo sapiens and Neanderthals, Neanderthals have never really become extinct. There is, however, 'evidence [that] shows that interaction of diverse forms is the source of more complex ones [...] mixing does not mean the eventual extinction of one in favour of the other; it simply results in transformation of the ontogenic system as the two forms converge' (Malasse et al., 1992: 50). Arguably, the human species may never have truly rid itself of another (humanoid) species (its DNA), like the Neanderthal.

Nevertheless, what the Neanderthal extinction case does show is how the flourishing of the human species that is capable of precise thought and action, brings with it total annihilation of other (humanoid) species. Being precise and efficient is thus one of our primal features that caused extinction of another humanoid species—*that* is our species' criminogenesis. Ever since, the human species has evolved into the dominating 'global species' of the planet and has 'become, through labor and technique, powerful "geological agents" [...] who are actively shaping the Earth system in ways that seriously undermine our safe spaces' (Harrington & Shearing, 2016: 13).

3.3 Dehumanisation, Total Annihilation and Exploitation

Our species' criminogenesis has incarnated our tendency to totally annihilate; not just externally toward other (humanoid) species, but also—perhaps especially—inwardly directed, as we have always totally exploited and annihilated our own species. Genocidal thinking and killing has occurred throughout our history and in every human community (Kiernan, 2008). Our agricultural pursuits comprised mass murder—a type of annihilative expansionism through the seizure of territory and the killing of its inhabitants (id. 161). From Neanderthals to ancient Greek Spartans, from colonisation to the Holocaust and the Russian war in Ukraine—in one way or another, wherever and whenever humans exist, total annihilation and exploitation seem to follow.

However, killing one's own species—called intraspecies killing—is not limited to human beings; other species also engage in (mass) killing of their own kind, like the most murderous meerkats, of which approximately 20% are killed by other meerkats (Goméz et al., 2016). In fact, most animals kill their fellow species to gain access to resources and territory, especially through infanticide (id.). Humans commit infanticide too, for different social, cultural, economic, and psychological reasons, the primary reason being unwanted pregnancies. Unwanted because of the huge economic impact it can have on a parent's life (Dongarwar, 2022). In other words, as in other species, infanticide is also caused by fear of one's resources being threatened.

What is different from other species is that humans specifically commit (mass) murder against human adults (not just children) also out of jealousy and hatred toward those whom we imagine to be different (Wrangham, 2004). We make a distinction in our intraspecies killing, caused by our reflexive consciousness that makes us have a conception of ourselves as a distinct being with higher-order desires, whereas other species with simple consciousness are unable to conceive their own distinctiveness (Jamieson, 1983). Other species do not (and cannot) kill out of distinction, which makes the human species truly exceptional when it comes to killing adults, as Harvard biological anthropologist Richard Wrangham argues (Castro, 2017). We kill and annihilate our own species because of making distinctions between ourselves.

Making distinctions is a mental skill the human species has attained by having been neuronally evolved into precision- and efficiency-thinking beings. We are able to know and reason with one another—*being sapient*—which makes 'human beings have an innate psychological readiness to see themselves in likeness to other human beings' and thus instead of 'homo sapiens [we] might better be called homo analogi' (Rudmin, 1993: 102). The human species' brains have neurons that are specialised in detailed facial recognition, reflected in how the face is the first thing we notice when looking at somebody. We are always looking for face-like stimuli and it is those stimuli we then value and devalue (cf. Bjornsdottir & Rule, 2016). This implies we are not only able to make a distinction between ourselves and other species; we also make a distinction between ourselves and other people in order to imagine other human beings as lesser beings or even as non-human. This is done because to consider a person as a non-human being goes against our human psyche to see likeness—to see analogues—in other humans. Therefore, to achieve total annihilation and commit genocide, we often need to dehumanise the other human being by imagining them as non-human, which allows us to distance ourselves from their likeness (cf. Arendt, 1994; Hagan & Rymond-Richmond, 2008; Haslam, 2019; Lang, 2017). Paradoxically, it is uniquely human to imagine other human beings as non-human; it is uniquely human to dehumanise.

In a similar vein, such all too human dehumanisation also allows the human species to enslave its own members, ripping away once more the humanity of another human beings in order to own "it". Dehumanising someone into a slave is an imaginative thinking exercise that has led to justification of slavery in philosophical ideas, laws and other types of codifications—all made possible, once more, by our precise reasoning and actions. It is well established that Greek philosophers justified slavery and the exploitation of a person by another human being (Hunt, 2017). For example, Platonic and Attic laws legitimated the existence of slaves, by reasoning that some people were born into slavery and that prisoners of war had to provide life-long service to their master. Aristotle's theorisation of unnatural slavery implies, for example, that those enslaved are inferior due to their lack of being able to reason (Gallagher, 2014), whereas free men and women are those that can reason.

Another paradox: it is our ability to imagine and reason that enables us to believe that another human being cannot reason and therefore can be subjected to slavery. However, this reasoned belief is anything but reasonable.

Not just by Greek philosophers; dehumanising human beings into inferiority, as expendable and exterminable "creatures", has filtered through in countless ideas about ownership of private property since the rise of agriculture and the fall of tribal communities (De Beauvoir, 1952). Later on, during the Egyptian colonial expansion into Africa between 3200 and 1200 BCE, for example, mass exploitation and massacres occurred (Adams, 1984). A few centuries after that, the Roman empire dehumanised its enemies into "barbarians" (Ferris, 2000), starting exploitation and genocidal campaigns against them (Kiernan, 2004), like the three-year siege of Carthage that resulted in the deaths of at least 150,000 Carthaginians and the enslavement of the 55,000 survivors (id.). The destruction of Carthage and the deaths of so many of its inhabitants...

...shared more modern features with recent tragedies such as the Armenian genocide, the Holocaust, and the Cambodian and Rwandan catastrophes. The perpetrators of these 20th-century crimes, like Cato, were pre-occupied with militaristic expansionism, the idealization of cultivation, notions of gender and social hierarchy, and racial or cultural prejudices. (id. 27)

It shows that racial or cultural prejudices are the very initiators as well as outcomes of dehumanisation, leading up to atrocious violence and exploitation throughout history, anywhere, with the common denominator being the human species.

European colonialism has been regarded as one of the most significant mass dehumanising exploitations and largest genocidal epochs the human species has ever managed to bring down upon its own members (cf. Crook et al., 2018; Wolfe, 2006). On 19 December 2022, Dutch Prime Minister Mark Rutte apologised for Dutch colonialism on behalf of the Dutch government:

For a long time I thought that the Netherlands' role in slavery was a thing of the past, something we had put behind us. But I was wrong. Centuries of oppression and exploitation still have an effect to this very day. In racist stereotypes. In discriminatory patterns patterns of exclusion. In social inequality. And to break those patterns, we also have to face up to the past, openly and honestly. [...] Today, on behalf of the Dutch government, I apologise for the past actions of the Dutch State: to enslaved people in the past, everywhere in the world, who suffered as a consequence of those actions, as well as to their daughters and sons, and to all their descendants, up to the present day. (Government of the Netherlands, 2022—online source)

Whenever Dutch slavery is discussed, the East India Company (EIC) is considered the main actor. However, rarely is the West India Company (WIC), mentioned and how they engaged in genocidal violence during the colonisation in "the West", present-day New York. Dutch merchant troops committed the intentional mass killing of native American members of the Tappan and Wecquaesgeek tribes during the Pavonia Massacre on 26 February 1643 (Axelrod, 2008). In recent decades, more and more knowledge has been built up about how European colonisation and the slave trade were genocidal and, in extreme cases, led to the complete extermination of peoples (Abugre, 2008; Best, 2009; Craemer,

2018; Drescher, 2018; Melby & Jones, 2008; Sartre, 1968; Thornton, 1987; Woolford & Hounslow, 2021).

The WIC colonisation was genocidal, arising from an (amoral) business model, in which profit-making prevailed over human dignity (cf. Abugre, 2008). Whereas genocidal and colonial violence may have decreased, the mindset of profit over people has got stuck. The Dutch mercantile mindset ("the EIC mentality") ensured a justification for dehumanising native Americans into "wildlings" that could be exterminated (Axelrod, 2008).

Either way, and whether it is the Dutch EIC and WIC genocidal colonisation, or the Congolese genocide by coloniser Belgium at the end of the nineteenth century, or the Herero genocide by the Germans, it becomes clear that dehumanisation is ingrained in European colonisation and in the laws and political-economic thought we used to justify colonisation (Mattei & Nader, 2008). Dehumanisation also underpins 18th-century racial theories that classified human beings based on skin colour and facial features (Edgar & Sedgewick, 1999), as well as that almost all of our world religions, including Islam, Judaism and Christianity, contain justifications to dehumanise others and enslave them (Ramelli, 2016) and/or massacre them (cf. Earl, 2011; Prior, 2006; Wright, 2022). The Enlightenment, as a reaction to absolutist religious monarchs and the inequalities and brutalities they inflicted upon their loyal subjects and enemies, also served as the beginning of modern Europe. However, the same Enlightenment of rationale and reason also enabled the dehumanisation of millions of people through the antisemitic Nuremberg Race Laws, leading to the Holocaust (Bauman, 2000; Horkheimer & Adorno, 1997). German criminologists actually undertook studies that supported the rationale and reason behind dehumanisation by the Nazi regime of the Jewish population that was eventually codified into the Nuremberg Race Laws (Rafter, 2008). Total annihilation and exploitation are always, everywhere, human-made, as well as always the outcomes of imaginative dehumanisation and of ideas justifying dehumanisation. Hence, the capability to imagine one's own species as lesser or as not part of that species (in this case, dehumanisation), often resulting in totally annihilating and exploitative tendencies, is one of the most characteristic aspects of being human.

3.4   CONCLUSION

This chapter posits that as a species, we are born from violence and mass harm, inflicted upon another human-like species, which resulted in the extinction of the Neanderthals. Ever since, total annihilation and exploitation through genocide and colonisation are specifically characteristic of the human species. Only the human species can imaginatively dehumanise and materialise such dehumanisation into mass exploitation (slavery) and total annihilation (repetitive genocides), even in modern times. It is *always* again, not *never* again. Eventually, it is the human imagination, enabled by our precision-focused brain and hands, that allows for such mass intraspecies exploitation and murder. It is the type of violence that we have unleashed upon our home planet as well.

REFERENCES

Abugre, C. (2008). Behind most mass violence lurk economic interests. In H. Melby & J. Y. Jones (Eds.), *Revisiting the heart of darkness—Explorations into genocide and other forms of mass violence* (pp. 273–280). Routledge.

Adams, W. Y. (1984). The first colonial empire: Egypt in Nubia, 3200–1200 BC. *Comparative Studies in Society and History, 26*(1), 36–71.

Arendt, H. (1994). On the nature of totalitarianism: An essay in understanding. In J. Kohn (Ed.), *Essays in understanding: 1930–1954* (pp. 328–360). Schocken Books.

Arendt, H. (2006). *Eichmann in Jerusalem.* Penguin.

Axelrod, A. (2008). *Profiles in folly: History's worst decisions and why they went wrong.* Sterling Publishing Company Inc.

Azevedo, F. A., Carvalho, L. R., Grinberg, L. T., Farfel, J. M., Ferretti, R. E., Leite, R. E., Filho, W. J., Lent, R., & Herculano-Houzel, S. (2009, April 10). Equal numbers of neuronal and nonneuronal cells make the human brain an isometrically scaled-up primate brain. *Journal of Comparative Neurology, 513*(5), 532–541.

Banks, W. E., d'Errico, F., Peterson, A. T., Kageyama, M., Sima, A., & Sánchez-Goñi, M. F. (2008). Neanderthal extinction by competitive exclusion. *PLoS ONE, 3*(12), 1–8.

Barakat, M. J., Field, J., & Taylor, J. (2013). The range of movement of the thumb. *The Hand, 8,* 179–182.

Bauman, Z. (2000). *Modernity and the holocaust.* Cornell University Press.

Beaulieu-Laroche, L., Brown, N. J., Hansen, M., Toloza, E. H., Sharma, J., Williams, Z. M., Frosch, M. P., Cosgrove, G. R., Cash, S. S., & Harnett, M.

T. (2021). Allometric rules for mammalian cortical layer 5 neuron biophysics. *Nature, 600*(7888), 274–278.

Best, S. (2009). Globalization of the human empire. In S. Dasgupta & J. N. Pieterse (Eds.), *Politics of globalization* (pp. 288–312). Sage.

Bezo, B., & Maggi, S. (2015). Living in "survival mode:" Intergenerational transmission of trauma from the Holodomor genocide of 1932–1933 in Ukraine. *Social Science & Medicine, 134*, 87–94.

Bjornsdottir, R. T., & Rule, N. O. (2017). The visibility of social class from facial cues. *Journal of Personality and Social Psychology, 113*(4), 530.

Browning, C. R. (1992). *Ordinary men: Reserve police battalion 101 and the final solution in Poland.* HarperCollins.

Castro, J. (2017). Do animals murder each other. https://www.livescience.com/60431-do-animals-murder-each-other.html

Caygill, H. (2016). Bataille and the Neanderthal extinction. *Georges Bataille and Contemporary Thought* (pp. 239–264).

Craemer, T. (2018). Comparative analysis of reparations for the holocaust and for the transatlantic slave trade. *The Review of Black Political Economy, 45*(4), 299–324.

Crook, M., Short, D., & South, N. (2018). Ecocide, genocide, capitalism and colonialism: Consequences for indigenous peoples and glocal ecosystems environments. *Theoretical Criminology, 22*(3), 298–317.

Curry, A. (2023). Neanderthals lived in groups big enough to eat giant elephants. https://www.science.org/content/article/neanderthals-lived-groups-big-enough-eat-giant-elephants

De Beauvoir, S. (1952). *The second sex.* Vintage.

Diamond, J. (2013). *The rise and fall of the third chimpanzee.* Random House.

Dongarwar, D. (2022). Extreme acts of violence: Infanticide and associated social constructs. *Handbook of anger, aggression, and violence* (pp. 1–17). Springer International Publishing.

Donoghue, D. (1979). Review of Hannah Arendt's *The Life of the Mind. The Hudson Review, 32*(2), 281–288.

Douglas, A. (2022). 10 animals that kill the most humans. https://www.worldatlas.com/animals/10-animals-that-kill-the-most-humans.html

Drescher, S. (2018). The Atlantic slave trade and the holocaust: A comparative analysis. In A. S. Rosenbaum (Ed.), *Is the Holocaust unique?: Perspectives on comparative genocide* (pp. 103–124). Routledge.

Earl, D. S. (2011). The Joshua delusion: Rethinking genocide in the Bible. *The Joshua Delusion* (pp. 1–190).

Edgar, A., & Sedgewick, P. (1999). *Cultural theory: The key concepts.* Routlege.

Eski, Y. (2020). An existentialist victimology of genocide? In Y. Eski (Ed.), *Genocide and victimology* (pp. 6–22). Routledge.

Eski, Y. (2022a). Omnia cadunt. Naar een victimologische verbeelding van onze vergankelijkheid Omnia cadunt (Translated from Toward a victimological imagination of our transience). *Tijdschrift over Cultuur & Criminaliteit, 12*(1), 58–71.

Eski, Y. (2022b). *A criminological biography of an arms dealer*. Routledge.

Ferris, I. (2000). *Enemies of Rome: Barbarians through Roman eyes*. Sutton Publishing.

Gallagher, W. (2014). Platonic and attic laws on slavery. *The Compass, 1*(1), 1–4.

Galway-Witham, J., Cole, J., & Stringer, C. (2019). Aspects of human physical and behavioural evolution during the last 1 million years. *Journal of Quaternary Science, 34*(6), 355–378.

Gates, B. (2014). The deadliest animal in the world. *Mosquito Week. The Gates Notes LLC*.

Gaudzinski-Windheuser, S., Kindler, L., MacDonald, K., & Roebroeks, W. (2023). Hunting and processing of straight-tusked elephants 125.000 years ago: Implications for Neanderthal behavior. *Science Advances, 9*(5), eadd8186.

Golding, W. (1955). *The inheritors*. Faber & Faber.

Gómez, J. M., Verdú, M., González-Megías, A., & Méndez, M. (2016). The phylogenetic roots of human lethal violence. *Nature, 538*(7624), 233–237.

Government of the Netherlands. (2022). Speech by Prime Minister Mark Rutte about the role of the Netherlands in the history of slavery. https://www.gov ernment.nl/documents/speeches/2022/12/19/speech-by-prime-minister-mark-rutte-about-the-role-of-the-netherlands-in-the-history-of-slavery

Greenbaum, G., Getz, W. M., Rosenberg, N. A., Feldman, M. W., Hovers, E., & Kolodny, O. (2019). Disease transmission and introgression can explain the long-lasting contact zone of modern humans and Neanderthals. *Nature Communications, 10*(1), 5003.

Hagan, J., & Rymond-Richmond, W. (2008). The collective dynamics of racial dehumanization and genocidal victimization in Darfur. *American Sociological Review, 73*(6), 875–902.

Harrington, C., & Shearing, C. (2016). *Security in the Anthropocene: Reflections on safety and care*. Transcript Verlag.

Haslam, N. (2019). The many roles of dehumanization in genocide. In L. S. Newman (Ed.), *Confronting humanity at its worst: Social psychological perspectives on genocide* (pp. 199–139). OUP.

Horkheimer, M., & Adorno, T. W. (1997). *Dialectic of enlightenment*. Verso.

Hublin, J. J. (2012). The earliest modern human colonization of Europe. *Proceedings of the National Academy of Sciences, 109*(34), 13471–13472.

Hunt, P. (2017). *Ancient Greek and Roman slavery*. John Wiley & Sons.

Jamieson, D. (1983). Killing persons and other beings. *Ethics and animals* (pp. 135–146).

Jones, A. (2016). *Genocide: A comprehensive introduction.* Routledge.

Katz, F. E. (1993). *Ordinary people and extraordinary evil; a Report on the beguilings of evil.* State University of New York Press.

Kiernan, B. (2008). *Blood and soil: A world history of genocide and extermination from Sparta to Darfur.* Yale University Press.

Kiernan, B. (2004). The first genocide: Carthage, 146 BC. *Diogenes, 51*(3), 27–39.

Kochi, T., & Ordan, N. (2008). An argument for the global suicide of humanity. *Borderlands, 7*(3), 1–21.

Kumar, A. (2012). The pollex-index complex and the kinetics of opposition. *Journal of Health and Allied Sciences NU, 2*(04), 80–87.

Lang, J. (2017). Explaining genocide: Hannah Arendt and the social-scientific concept of dehumanization. *The Anthem Companion to Hannah Arendt, 187,* 188.

Lombard, M., & Phillipson, L. (2010). Indications of bow and stone-tipped arrow use 64,000 years ago in KwaZulu-Natal, South Africa. *Antiquity, 84*(325), 635–648.

Malasse, A. D., Debénath, A., & Pelegrin, J. (1992). On new models for the Neanderthal debate. *Current Anthropology, 33*(1), 49–54.

Mattei, U., & Nader, L. (2008). *Plunder: When the rule of law is illegal.* John Wiley & Sons.

Melby, H., & Jones, J. Y. (2008). *Revisiting the heart of darkness—Explorations into genocide and other forms of mass violence.* Routledge.

Metz, L., Lewis, J. E., & Slimak, L. (2023). Bow-and-arrow, technology of the first modern humans in Europe 54,000 years ago at Mandrin, France. *Science Advances, 9*(8), eadd4675.

Milgram, S. (1963). Behavioral study of obedience. *Journal of Abnormal and Social Psychology, 67,* 371–378.

Napier, J. R. (1956). The prehensile movements of the human hand. *The Journal of bone and joint surgery. British volume, 38*(4), 902–913.

Prior, J. M. (2006). 'Power' and 'the Other' in Joshua: The brutal birthing of a group identity. *Mission Studies, 23*(1), 27–43.

Rafter, N. (2008). Criminology's darkest hour: Biocriminology in Nazi Germany. *Australian & New Zealand Journal of Criminology, 41*(2), 287–306.

Ramelli, I. (2016). *Social justice and the legitimacy of slavery: The role of philosophical asceticism from ancient Judaism to late antiquity.* OUP.

Rolian, C. (2016). The role of genes and development in the evolution of the primate hand. *The Evolution of the Primate Hand: Anatomical, Developmental, Functional, and Paleontological Evidence* (pp. 101–130).

Rudmin, F. W. (1993). Review of *The Genocidal Mentality,* by R. J. Lifton & E. Markusen. *Peace Research, 25*(4), 102–104.

Sartre, J. P. (1968). Genocide. *New Left Review, 48,* 13–25.

Silas SSF, B. S. (1997). Searching for the other. *Reviews in Religion & Theology, 4*(4): 74–79.

Smith, F. A., Elliott-Smith, R. E., Lyons, S. K., & Payne, J. L. (2018). Body size downgrading of mammals over the late Quaternary. *Science, 360*(6386), 310–313.

Solís, O. R. (2004). Some thoughts on the state of the world from a transactional analysis perspective. *Transactional Analysis Journal, 34*(4), 341–346.

Thornton, R. (1987). *American Indian holocaust and survival: A population history since 1492.* University of Oklahoma Press.

Troxell, E. L. (1936). The thumb of man. *The Scientific Monthly, 43*(2), 148–150.

Walsh, E. A. (2010). *Analogy's territories: Ethics and aesthetics in darwinism, modernism, and cybernetics.* University of California.

Weck, O. L. de. (2022). *Technology roadmapping and development: A quantitative approach to the management of technology.* Springer Nature.

Wells, P. S. (2011). The iron age. *European prehistory: A survey* (pp. 405–460).

Williams, L. (2022). 10 deadliest animals to humans. https://www.discoverwild life.com/animal-facts/deadliest-animals-to-humans/

Wolfe, P. (2006). Settler colonialism and the elimination of the native. *Journal of Genocide Research, 8*(4), 387–409.

Woolford, A., & Hounslow, W. (2021). Symbiotic victimisation and destruction: Law and human/other-than-human relationality in genocide. In Y. Eski (Ed.), *Genocide & victimology* (pp. 86–101). Routledge.

Wrangham, R. (2004). Killer species. *Daedalus, 133*(4), 25–35.

Wright, D. C. S. (2022). Arab Colonialism and the roots of the Golden Age of Islam. https://papers.ssrn.com/sol3/papers.cfm?abstract_id=4020040

Young, R. W. (2003). Evolution of the human hand: The role of throwing and clubbing. *Journal of Anatomy, 202*(1), 165–174.

CHAPTER 4

We, the Planet-Eating People: The Anthropocene Extinction and Omnicide

Abstract This chapter imagines the evolution of the human species into an omnicidal species, given the unique characteristic of the human species to bring about global-scale destruction and the sixth mass extinction. It argues that the human imagination of being a superior and eternal species on Earth has itself contributed to this destruction. The stubborn belief that we will overcome environmental problems through technology is hindering us from addressing the issue of mass extinction realistically. Imaginably, the human species is capable of causing, accelerating, and at the same time denying its own extinction.

Keywords Anthropocene mass extinction · Climate change · Omnicide · Extinction denial

4.1 Human Uniqueness and Environmental Annihilation

I'd like to share a revelation that I've had during my time here. It came to me when I tried to classify your species and I realized that you're not actually mammals. Every mammal on this planet instinctively develops a

Y. Eski, *A Criminology of the Human Species*, Palgrave Studies in Green Criminology, https://doi.org/10.1007/978-3-031-36092-3_4

natural equilibrium with the surrounding environment but you humans do not. You move to an area and you multiply and multiply until every natural resource is consumed and the only way you can survive is to spread to another area. There is another organism on this planet that follows the same pattern. Do you know what it is? A virus. Human beings are a disease, a cancer of this planet. You're a plague and we are the cure. (Agent Smith to Morpheus in *The Matrix*, 1999)

It has been argued (Hird, 2002: 103) that Agent Smith—a software programme in the movie *The Matrix*, designed to control the avatars of real human beings—fears the nearly unique propensity of the human species to 'shorten all food chains in the web, eliminate most intermediaries and focus all biomass on themselves. Whenever an outside species tries to insert itself into one of these chains, to start the process of complexication again, it is ruthlessly expunged as a "weed"' (De Landa, 2000: 108). What if Agent Smith is right?

Throughout human history, the human species has managed to acquire control over nutrient cycles in harmony with other species: 'cattle and certain crops went hand in hand: the manure of the cattle, which were raised on cereals, could be plugged back into the system as fertilizer, closing the nutrient cycle' (id. 122). Human beings lived in harmony with and in between ecosystems. However, due to complexification through industrialisation and the consumer society, we have irreversibly disrupted highly complex systems of all kinds of species and other (micro)flora and fauna by sterilising the lands and soils all life forms depend on (id.). As criminologist Clifford Shearing stated, 'we must now be conceived of as integral to earth systems [...] as biophysical "actants" who have, through our actions, significantly reshaped the earth' (Shearing, 2015: 257). Instead of being part of ecosystems, we took full control over the ecosystemic cycles themselves, determining the growth of all things we consume (De Landa, 2000: 122).

What is problematic is that our way of life assumes that unlimited growth is good and, therefore, a just purpose to live by, despite evidence to the contrary since the 1970s (Wagar, 1970). Growth for growth's sake is disastrous (Deans & Larson, 2008); in fact, 'growth for the sake of growth is the ideology of the cancer cell' (id.). Similarly, Julian Huxley, the brother of Aldous Huxley (who wrote *Brave New World* in 1932) and an evolutionary biologist, once claimed that 'the human race will be the cancer of this planet' (n.d.).

Perhaps the human species has become a cancer of Earth's biosphere and all that lives within it, where we function as 'unrestricted autotelic growth [...] that uses life forces for no other end than that of maintaining itself [...] the phenomenon of excessive purposeless vitality' (Heinämaa, 2018: 58). If so, as a species then, we are actually not a "being for-itself", living a conscious life with a purposive nature, but a "being in-itself", living a self-sufficient, contingent life (Sartre, 1943), a 'surplus of life, [not] structured by purpose' (Kolnai, 1929: 72 in Heinämaa, 2018: 54).

Despite such purposeless life, the human species believes it is destined to do great things and to be "for-itself", forever. We imagine ourselves to be unique and virtuous among all creatures, given our free will, individual autonomy and the independent cognitive decisions we make that are supposedly based on reason and free moral choice (cf. Descartes, 1641; Hume, 1740; Kant, 1781; Spinoza, 1677). These virtues of human dignity are acknowledged in the Charter of the United Nations, the Universal Declaration of Human Rights (1948) and it is precisely that human dignity that would distinguish the human species from other species.

However, our assumed unique traits, behaviours, and abilities, in particular with regard to cognitive abilities, are prevalent in other species, like carrion crows, for example, that display 'a neuronal response in the palliative end brain during the performance of a task that correlates with their perception of a stimulus', an activity that implies consciousness (Nieder et al., 2020: 1626). So, while it is a common belief that humans are the only conscious species on Earth, other creatures do exhibit signs of consciousness through their behaviours and brain activity, and while we may perceive ourselves as the superior beings in terms of consciousness, it is important to acknowledge that we are not alone in this regard.

However, what makes us truly unique is that, because of human-made climate change, the human species violates human dignity *and* that of a wide range of species, as animals have dignity too (cf. Manfred Nowak in Kaufmann et al., 2010). We are the only species that daily destroys other species and robs them of their dignity on a mass scale, assisted by modern science and technology. That human-based harm inflicted upon the ecosystem itself can be considered a form of total exploitation and annihilation of all flora and fauna, referred to as ecocide (White, 2017: 60). The difference with genocide and colonisation, however, is that we do not have to dehumanise nature to distance ourselves from it in order to destroy it. In 1637, philosopher René Descartes famously declared, 'I think, therefore I am', which laid the foundation for modern philosophy and rationalism. However, this idea also created a perceived separation between humans and the natural world, leading to a disconnection and dispiritedness toward animals and nature as a whole (Crist, 2013). And we are having a hard time reconnecting with nature, if only because we are destroying it every day.

Such daily ecocide, leading up to the sixth mass extinction, is almost entirely self-inflicted, while also being denied. The alarming conclusion by the UN Intergovernmental Panel on Climate Change (IPCC, 2021) testifies to it. Rapid climate change is causing our planet to heat up irreversibly and is going to result in an uninhabitable planet. One would expect that this would give a global sense of urgency, and lead to a complete mental and lifestyle change of the human species. That is, however, not the case (Lees et al., 2020).

It was already shown in the 1970s (Meadows et al., 1972) that our population growth and way of life could lead to the end of human civilisation. It has been known for half a century that humans are responsible for climate change and the call for awareness of this is about the same age (Nakicenovic et al., 2000). Climate change is a sign of the sixth mass extinction that started 8,000 years ago and is currently taking place (Koh, 2004). It concerns the Anthropocene (sometimes called Holocene) mass extinction, in which approximately one million species of animals, insects and plants are now at risk of extinction.

The difference with previous mass extinctions (see Chapter 2) is that it is not natural forces but one living species that threatens all biodiversity with accelerated mass extinction: the human species. Whereas past extinctions were caused by uncontrollable and geological events, the current sixth mass extinction is caused by humans. It is because of us that the current mass extinction has a biotic cause—us—and not a physical one, like volcanism or comets. Our impact on the planet, though, is similar to the Cretaceous cometary collision in terms of the physical changes caused (Eldredge, 2001; Naggs, 2017).

Let us consider the sixth mass extinction in more detail. The literature disagrees on how many phases of the Anthropocene mass extinction can be identified. Eldredge (2001) speaks of two phases, one beginning 100,000 years ago with human dispersal, and the follow-up phase beginning 10,000 years ago when we discovered agriculture. However, Avise et al. (2008) speak of three phases of 'the biotic holocaust [that] is due—directly or indirectly—to local, regional, and global environ-mental impacts from a burgeoning human population' (id. 11,453). Phase 1 started about 50,000–100,000 years ago, when homo sapiens began spreading across the globe (id.). Recent archaeological research showed that other, earlier humanoid species, like the Neanderthal, may already have left their first ecological footprints shaping the landscape through the use of fire at least 400,000 years ago (Roebroeks et al., 2021). Since then, the human species has followed in their footsteps and about 10,000 years ago, phase 2 commenced; human population numbers increased and agri-cultural land-use grew. Phase 3, starting with the industrial revolution in the eighteenth century, brought about environmental changes and loss of biodiversity (Avise et al., 2008).

Ever since, as has been estimated, we have been losing 0.25 per cent of remaining species annually, meaning that on average 12,000 species die out per year (Wilson, 1993). It reveals the magnitude of the sixth mass extinction in which countless species are being decimated by the human species (Avise et al., 2008). So, despite us being 'clearly a species of animal', we are 'however, behaviourally and ecologically peculiar an animal' (Eldredge, 2001: 2). As paleoecologist Cindy Looy pointed out: 'the human race is a geological force' (Looy in Waarlo, 2022—online source). In having become a force of nature ourselves, the human species has crossed four of the nine biospheric boundaries, according to Harris et al. (2019):

- Violation of the boundary of the biome through deforestation;
- Violation of the border of biodiversity through, among other things, over-fishing and hunting;
- Violating the limit of the hydrological circulation whereby there is an increase in phosphorus in rivers and seas, culminating in ocean acidification and underwater dead zones and;
- Violating the atmospheric limit due to CO_2 emissions.

The current mass extinction is amplified and accelerated by virus outbreaks (cf. Chowdhury et al., 2021) and environmental pollution (Villarrubia-Gómez et al., 2018). If we continue this pattern, in about 400 years we will be living at the peak of the Anthropocene mass extinc-tion (Barnosky et al., 2011). Our greenhouse gas emissions in particular are the main culprit, changing the climate just as quickly as (and perhaps faster than) during the third mass extinction (Greshko, 2019). If the human species continues on its current trajectory, life as we know it will become an agonising descent into irreversible damage to our ecosystems, leading to the near-certain destruction of nearly all existing biodiversity, including ourselves (Barnosky et al., 2011; Leslie, 2002). And despite all these clear indicators and research, mass extinction denial exists.

4.2   Denying the Anthropocene Mass Extinction

Whenever we imagine climate change, we think of more rain and more heat, but not so much that: the Amazon region emits more CO_2 than it absorbs (Gatti et al., 2021); there are more and more "dead zones" in the oceans where oxygen is lacking, and flora and fauna are becoming extinct below the waterline (Breitburg et al., 2018); or mega-droughts are becoming more common and can wipe out entire civilisations (Garreaud et al., 2021). Imaginatively, the human species does not consider its own extinction due to climate change.

As a matter of fact, although we consider ourselves climate change victims, 'we are also the ones who can do something about it' (Kaag, 2021—online source). The imagination that we can still do something about climate change reflects a belief in the eternal survival of humanity; a belief that stems from our survival instinct as a species. The assumption of the eternity of man as a species is of all times. Ancient Greek philosophers (Simpson, 1985), Christians (Joyce, 2020), liberal philosophers (Paik, 2020) and Marxists (Kerr, 2002); they all talk about the reversibility of the 'Apocalypse' and the eternal survival of man. Doing something about climate change indicates a strong, inexhaustible will to survive that nowadays is accompanied by a denial of extinction (Lees et al., 2020).

That disbelief regarding our end as a species is ingrained in climate policy, where the assumption about human survival is not questioned. The awareness of being global climate victims is accompanied by increased risk awareness and intensive reflection on problem solving (Mythen & McGowan, 2017), which translates into an international mobilisation against climate change. In that mobilisation, politics, science and the public debate focus on the survival of the human species and that the climate crisis can be managed. Whether it is neoliberal social engineering or environmental movements that are convinced that climate change control and reversal is necessary and possible: across the entire ideological-political spectrum, a stubborn belief is then perpetuated that we will (have to) survive anyway (Chinn et al., 2020).

That stubborn belief seems to make us forget about the previous five mass extinctions; they form evidence that we too might become extinct one day, yet mass extinction is harder to consider and debate over than ever before (Lees et al., 2020). It leaves behind a troubling observation: even as we hurtle toward a potential end of our species and countless others, there is a growing stubbornness in our belief that we can survive and a wilful disbelief in our own possible extinction (Eski, 2022).

This is, however, understandable. The prospect of extinction contradicts the vision of everlasting human progress both on and beyond Earth, which further contributes to denial of human extinction (Cowie et al., 2022). Similar to our individual fear of death, which we tend to avoid and deny in modern society, human extinction denial has become deeply rooted in our sociobiological makeup (Lees et al., 2020). In fact, it resembles the same fear and dread that burden us daily, but now on a much larger scale. As the late sociologist Zygmunt Bauman (2006) noted, nowadays we are more inclined to avoid and deny the reality of death than ever before in history. Living with the fear of extinction and then becoming extinct is a way of life that does not fit into the twenty-first century of the everyday global interconnectedness, of climate change awareness, or of the highly advanced technological progress, through which we as a species imagine ourselves (online) to be an immortal species (cf. Lockwood, 2011; Popa, 2021). Our urge to survive has, in a sense, become an existential exercise through which we culturally embed ourselves in our existence as an 'eternal' species.

By approaching climate change as a problem that requires solving, we *again* place ourselves hierarchically above other forms of life; a position that is fundamentally misguided (Heise, 2016). Our attitude toward climate change reflects our perception of ourselves as an eternal species and our relationship with the Earth's biodiversity. We assume that we will continue to exist forever and that all other forms of life are subservient to our needs. The possibility of our own extinction, however, is hardly ever mentioned in public discourse or political debates (Heise, 2016; Lees et al., 2020). Instead, we focus solely on solving climate change for our own benefit (Popovski & Mundy, 2012), perpetuating our anthropocentric view of and place in the world (Janssen & Schuilenburg, 2021). The obsession with climate victimisation and its solutions blinds us to the fact that we, too, are susceptible to extinction, just like the countless species that disappeared during the previous five mass extinctions (Chapter 2).

Our denial of our mortality is evident in our approach to climate change solutions, which assume that we will inevitably survive as a species on Earth (Cohen, 2013). In addressing the climate crisis, we ought to adopt a cosmopolitan imagination that acknowledges our interconnectedness with the planet and its diverse life forms. This will require a radical transformation of our way of life, one that embraces global awareness and responsibility (Padilla, 2021: 163–164).

However, that call for 'intelligence' is again a reflection of the imagination that humankind will and must survive, an imagination that is materialised into climate-saving projects. Think of projects in which mega plants are grown that absorb CO_2 in Iceland (Von Strandmann et al., 2019), or superglacier blankets that are placed over ice plains in Switzerland (Hoelzle et al., 2011), or a global manufactured cloud cover that cools the Earth's surface (Diamond et al., 2020), or advanced cloud-seeding technology with which rain can be created to prevent future drought (Wang et al., 2022). If all of this does not work on the surface of the Earth, there are ideas for living underground (cf. Bobylev, 2006) or under water (cf. Ilardo et al., 2018), or in space and on other planets (cf. Kaku, 2018). Tesla and SpaceX owner Elon Musk wants to take humans to Mars to save the species and make us an interplanetary species, for example (Musk, 2017). Chapter 6 delves further into space as the final frontier to exploit and annihilate.

Like human beings, there are animals, especially mammals, that show awareness of (their own) death (Anderson, 2020; King, 2016; Pierce, 2012), being suicidal, even (Hediger, 2018; Peña-Guzmán, 2017). Yet, it seems to remain unknown whether animals are aware of (their own) mass extinction. The human species and its active denial of mass extinction, then, deviate from other sentient species, simply because we *are* denying that the sixth mass extinction is coming and is accelerated by us (Cowie et al., 2022; Lees et al., 2020; Waarlo, 2022). It is a denial ingrained in finding solutions to climate change that often serve as a cover-up for our denial of the possibility of our own extinction. The human species and its green aspirations only perpetuate the belief in eternal human survival and distance humankind further from the reality of the Anthropocene mass extinction. In essence, it is very human to be able to imagine the approaching end of Earth while denying our own extinction. It fashions the Anthropocene extinction into an all too human omnicide.

4.3   All-Too-Human Omnicide

That a single species—we, the planet-eating species—is causing, accelerating and denying its own mass extinction is novel compared to the previous mass extinctions. No other species ever walked this Earth that has singlehandedly caused a mass extinction. Like volcanoes, earthquakes and landslides, we have the potential to destroy our environment and entire planet beyond repair.

The past and current mass extinctions demonstrate that virtually all forms of biodiversity that existed on Earth have died out and have been replaced by new biodiversity. Animals and their ecosystems constitute a functional entity of which the human species is a part, and thus, the human-caused acceleration of the sixth mass extinction threatens not only human civilisation but also the survival of animals (Ceballos et al., 2017). In fact, we have always hastened the extinction of large animals, especially those posing a threat to humans and our livestock (Haynes, 2018), which, in turn, favoured human and livestock population growth (Criscuolo & Sueur, 2020). As a result, the human species dominates Earth as an omnicidal agent (Torres, 2018).

In geological history, Earth's terrestrial biosphere has always gradually changed its composition by means of sequential mass extinctions. It is expected to do so for a long time into the future (Pievani, 2014). There is a geological logic behind mass extinctions, ensuring an evolution of biodiversity (Courtillot & Gaudemer, 1996; Greshko, 2019). For example, the extinction of gigantic creatures, such as *Tyrannosaurus rex*, has allowed humans and other species of mammals to evolve. Such geological change, as shown earlier, causes mass extinction of life to enable mass rebirth and allows new life to roam Earth. Mass extinctions do not necessarily happen abruptly or unexpectedly either; they have no clear start or end point, nor is mass extinction all-defeating (Logan, 2018). Species from the time of the dinosaurs are still alive today, the crocodile and the turtle being the clearest examples (Lyson et al., 2019). Mass extinction thus encompasses a process that gradually begins for several (related or unrelated) causes (e.g., climate change, volcanism and/or meteorites) and does not necessarily lead to the complete end of all living beings. Mass extinction has its own logic whereby Earth can change its biodiversity throughout geological history, exchanging it for new and different biodiversity.

For us as human beings this means that we can be exchanged for other life or that we can evolve into another (human) being. To imagine ourselves to be the end station (in our current form) and ruler of all biodiversity on this planet is therefore questionable, if not naïve; it is most definitely human to imagine it. What can be stated with some certainty is that biodiversity needs a habitable planet anyway, but the reverse is not necessarily true: in other words, Earth does not need human life or any life to exist (forever) (Seager, 2013). To imagine the opposite, namely, that we must "save the planet" is, arguably, a false imagination.

Still, it is the human species that, in being the most intelligent life form on Earth, is simultaneously the most self-destructive species; a most violent tendency we ignore instead of actively imagine. Imaginably, if we do manage to survive and become an eternal (planet-eating) species, we may become a force of nature that has skilfully disabled Earth's capability to renew its surface with novel biodiversity. For the first time in geological history, a species—the human species—would be able to threaten Earth by immobilising its capacity to accommodate mass extinctions and mass rebirths. The human species may have the potential to disable Earth's way of reproducing life out of mass extinctions. Like cancer cells, the human species would replace the normal and healthy "cells" of Earth, continuing to divide, growing rapidly without maturing into useful elements of Earth, while evading the immune system of the planet and spreading to other parts of the "body" of planets, vis-à-vis outer space, forming an interplanetary, metastatic cancer. We would have eaten Earth, while eyeballing other planets.

4.4   Conclusion

This chapter imagined how the human species is an omnicidal being and that being omnicidal is uniquely human. By imagining ourselves as a unique species on Earth, we have claimed superiority over our natural environment and other species in it. In doing so, and next to being born of violence, humanity has set off and accelerated a global-scale violent destruction, namely, the sixth mass extinction, which is most likely irreversible. An irreversibility that is fully denied and if it is acknowledged, then a stubborn belief persists that we shall overcome it in one way or another through climate awareness, going green and environment-friendly smart-technological solutions. It is a stubborn belief that *amplifies* mass extinction denial. Hence, imaginably, we are a species (perhaps the only species) that is capable of *causing*, *accelerating* and *denying* its self-inflicted mass extinction (and that of other species).

REFERENCES

Anderson, J. R. (2020). Responses to death and dying: Primates and other mammals. *Primates, 61*(1), 1–7.

Avise, J. C., Hubbell, S. P., & Ayala, F. J. (2008). In the light of evolution II: Biodiversity and extinction. *Proceedings of the National Academy of Sciences of the United States of America, 105*, 11453–11457.

Barnosky, A. D., Matzke, N., Tomiya, S., Wogan, G. P., Swartz, B., Quental, T. B., Marshall, C., McGuire, J. L., Lindsey, E. L., Maguire, K. C., & Mersey, B. (2011). Has the Earth's sixth mass extinction already arrived? *Nature, 471*(7336), 51–57.

Bauman, Z. (2006). *Liquid fear*. Polity.

Bobylev, N. (2006). Strategic environmental assessment of urban underground infrastructure development policies. *Tunnelling and Underground Space Technology, 21*(3–4), 469.

Breitburg, D., Levin, L. A., Oschlies, A., Grégoire, M., Chavez, F. P., Conley, D. J., Garçon, V., Gilbert, D., Gutiérrez, D., Isensee, K., Jacinto, G. S. (2018). Declining oxygen in the global ocean and coastal waters. *Science, 359*(6371).

Ceballos, G., Ehrlich, P. R., & Dirzo, R. (2017). Biological annihilation via the ongoing sixth mass extinction signaled by vertebrate population losses and declines. *Proceedings of the National Academy of Sciences of the United States of America, 114*, E6089–E6096.

Chinn, S., Hart, P. S., & Soroka, S. (2020). Politicization and polarization in climate change news content, 1985–2017. *Science Communication, 42*(1), 112–129.

Chowdhury, R. B., Khan, A., Mahiat, T., Dutta, H., Tasmeea, T., Armaan, A. B., Fardu, F., Roy, B. B., Hossain, M. M., Khan, N. A., & Amin, A. N. (2021). Environmental externalities of the COVID-19 lockdown: Insights for sustainability planning in the Anthropocene. *Science of The Total Environment* (p. 147015).

Cohen, S. (2013). *States of denial: Knowing about atrocities and suffering*. John Wiley & Sons.

Courtillot, V., & Gaudemer, Y. (1996). Effects of mass extinctions on biodiversity. *Nature, 381*(6578), 146–148.

Cowie, R. H., Bouchet, P., & Fontaine, B. (2022). The sixth mass extinction: Fact, fiction or speculation? *Biological Reviews, 97*(2), 640–663.

Criscuolo, F., & Sueur, C. (2020). An evolutionary point of view of animal ethics. *Frontiers in Psychology, 11*, 403.

Crist, E. (2013). Ecocide and the extinction of animal minds. In M. Bekoff (Ed.), *Ignoring nature no more: The case for compassionate conservation* (pp. 45–61). University of Chicago Press.

De Landa, M. (2000). *A thousand years of nonlinear history*. Swerve Editions.

Deans, G., & Larson, M. (2008). Growth for growth's sake: A recipe for a potential disaster. *Ivey Business Journal, 72*, 1–12.

Descartes, R. (1641 [1988]). Meditations on first philosophy. In J. Cottingham, R. Stoothoff, & D. Murdoch (Eds.), *Descartes: Selected philosophical writings* (pp. 73–121). Cambridge University Press.

Diamond, M. S., Director, H. M., Eastman, R., Possner, A., & Wood, R. (2020). Substantial cloud brightening from shipping in subtropical low clouds. *AGU Advances, 1*(1), e2019AV000111.

Eldredge, N. (2001). The sixth extinction. https://www.biologicaldiversity.org/programs/population_and_sustainability/extinction/pdfs/Eldridge-sixth-extinction.pdf

Eski, Y. (2022). Omnia cadunt. Naar een victimologische verbeelding van onze vergankelijkheid Omnia cadunt (Translated from Toward a victimological imagination of our transience). *Tijdschrift over Cultuur & Criminaliteit, 12*(1), 58–71.

Garreaud, R. D., Clem, K., & Veloso, J. V. (2021). The South Pacific Pressure Trend Dipole and the Southern Blob. *Journal of Climate, 34*(18), 7661–7676.

Gatti, L. V., Basso, L. S., Miller, J. B., Gloor, M., Gatti Domingues, L., Cassol, H. L., Tejada, G., Aragão, L. E., Nobre, C., Peters, W., & Marani, L. (2021). Amazonia as a carbon source linked to deforestation and climate change. *Nature, 595*(7867), 388–393.

Greshko, M. (2019). Wat waren de 5 massa-uitstervings en wat veroorzaakte ze? https://www.nationalgeographic.nl/wetenschap/2019/09/wat-waren-de-5-massa-uitstervings-en-wat-veroorzaakte-ze

Harris, N., Goldman, E. D., & Gibbes, S. (2019). Spatial database of planted trees (SDPT) version 1.0. *Technical Note, World Resources Institute.*

Haynes, G. (2018). The evidence for human agency in the late pleistocene megafaunal extinctions. In D. A. DellaSala & M. I. Goldstein (Eds.), *The Encyclopedia of the Anthropocene* (pp. 219–226). Elsevier.

Hediger, R. (2018). Animal suicide and "anthropodenial." *Animal Sentience, 20*(16), 1–3.

Heinämaa, S. (2018). Strange vegetation: Emotional undercurrents of Tove Jansson's Moominvalley in November. *SATS, 19*(1), 41–67.

Heise, U. K. (2016). *Imagining extinction.* University of Chicago Press.

Hird, M. J. (2002). Re(pro)ducing sexual difference. *Parallax, 8*(4), 94–107.

Hoelzle, M., Darms, G., Lüthi, M. P., & Suter, S. (2011). Evidence of accelerated englacial warming in the Monte Rosa area, Switzerland/Italy. *The Cryosphere, 5*(1), 231–243.

Hume, D. (1740 [1978]). A Treatise of Human Nature, L. A. Selby-Bigge & P. H. Nidditch (Eds.), 2nd edition. OUP.

Ilardo, M. A., Moltke I., Korneliussen, T. S., Cheng, J., Stern, A. J., Racimo, F., de Barros Damgaard, P., Sikora, M., Seguin-Orlando, A., Rasmussen, S., & van den Munckhof, I. C. (2018). Physiological and genetic adaptations to diving in sea nomads. *Cell, 173*(3), 569–580.

IPCC. (2021). Climate change 2021: The physical science basis report. https://www.ipcc.ch/report/ar6/wg1/downloads/report/IPCC_AR6_WGI_SPM.pdf

Janssen, J., & Schuilenburg, M. (2021). Het antropoceen. De criminologische uitdaging in de 21ste eeuw. *Tijdschrift over Cultuur & Criminaliteit, 11*(1), 3–13.

Joyce, C. (2020). Responses to apocalypse: Early Christianity and extinction rebellion. *Religions, 11*(8), 384.

Kaag, S. (2021). LinkedIn post over klimaatverandering. https://www.linkedin.com/posts/sigrid-kaag_ipcc-ongepercentC3percentABvenaarde-klimaatveran dering-leidt-activity-6830433144768909312-UIs

Kaku, M. (2018). *The future of humanity: Terraforming Mars, interstellar travel, immortality, and our destiny beyond.* Penguin.

Kant, I. (1781[1999]). *Critique of Pure Reason,* P. Guyer, & A. W. Wood (Eds.). Cambridge University Press.

Kaufmann, P., Kuch, H., Neuhaeuser, C., & Webster, E. (2010). *Humiliation, degradation, dehumanization: Human dignity violated.* Springer.

Kerr, P. (2002). Saved from extinction: Evolutionary theorising, politics and the state. *The British Journal of Politics and International Relations, 4*(2), 330–358.

King, B. J. (2016). Animal mourning: Précis of How animals grieve. *Animal Sentience, 4*(1), 1–6.

Koh, L., et al. (2004). Species coextinctions and the biodiversity crisis. *Science, 305*(5690), 1632–1634.

Lees, A. C., Attwood, S., Barlow, J., & Phalan, B. (2020). Biodiversity scientists must fight the creeping rise of extinction denial. *Nature Ecology & Evolution, 4*(11), 1440–1443.

Leslie, J. (2002). *The end of the world: The science and ethics of human extinction.* Routledge.

Lockwood, G. M. (2011). Social egg freezing: The prospect of reproductive 'immortality' or a dangerous delusion? *Reproductive Biomedicine Online, 23*(3), 334–340.

Logan, R. K. (2018). The Anthropocene and climate change: An existential crisis. https://www.researchgate.net/profile/Robert_Logan5/publication/322437374_The_Anthropocene_and_Climate_Change_An_Existential_Cri sis/links/5a58cb73aca2727d60814d66/The-Anthropocene-and-Climate-Change-An Existential-Crisis

Lyson, T. R., Miller, I. M., Bercovici, A. D., Weissenburger, K., Fuentes, A. J., Clyde, W. C., Hgadorn, J. W., Butrim, M. J., Johnson, K. R., Fleming, R. F., & Barclay, R. S. (2019). Exceptional continental record of biotic recovery after the Cretaceous-Paleogene mass extinction. *Science, 366*(6468), 977–983.

Meadows, D., et al. (1972). *The limits to growth; A report for the club of Rome's project on the predicament of mankind.* Universe Books.

Musk, E. (2017). Making humans a multi-planetary species. *New Space, 5*(2), 46–61.

Mythen, G., & McGowan, W. (2017). Cultural victimology revisited: Synergies of risk, fear and resilience. *Handbook of victims and victimology* (pp. 364–378). Routledge.

Naggs, F. (2017). Saving living diversity in the face of the unstoppable sixth mass extinction: A call for urgent international action. *The journal of population and sustainability, 1*(2), 67–81.

Nakicenovic, N., Alcamo, J., Davis, G., Vries, B. D., Fenhann, J., Gaffin, S., Gregory, K., Grubler, A., Jung, T. Y., Kram, T., & La Rovere, E. L. (2000). Special report on emissions scenarios. https://escholarship.org/content/qt9 sz5p22f/qt9sz5p22f.pdf

Nieder, A., Wagener, L., & Rinnert, P. (2020). A neural correlate of sensory consciousness in a corvid bird. *Science, 369*(6511), 1626–1629.

Padilla, L. A. (2021). The Anthropocene: Are we in the midst of the sixth mass extinction? *Sustainable development in the Anthropocene* (pp. 93–167). Springer.

Paik, P. Y. (2022). Apocalypse and extinction. In C. Nikou (Ed.), *Imaginaires postapocalyptiques: Comment penser l'après.* UGA Éditions (pp. 221–240).

Peña-Guzmán, D. M. (2017). Can nonhuman animals commit suicide? *Animal Sentience, 2*(20), 1–24.

Pierce, J. (2014). *The last walk: Reflections on our pets at the end of their lives.* University of Chicago Press.

Pievani, T. (2014). The sixth mass extinction: Anthropocene and the human impact on biodiversity. *Rendiconti Lincei, 25*(1), 85–93.

Popa, C. (2021). Online afterlives: Immortality, memory, and grief in digital culture. *Metacritic Journal for Comparative Studies and Theory, 7*(1), 301–306.

Popovski, V., & Mundy, K. G. (2012). Defining climate-change victims. *Sustainability Science, 7*(1), 5–16.

Roebroeks, W., MacDonald, K., Scherjon, F., Bakels, C., Kindler, L., Nikulina, A., Pop, E., & Gaudzinski-Windheuser, S. (2021). Landscape modification by Last Interglacial Neanderthals. *Science Advances, 7*(51), 1–13.

Sartre, J. P. (1943). *L'Être et le néant : Essai d'ontologie phénoménologique.* Éditions Gallimard.

Seager, S. (2013). Exoplanet habitability. *Science, 340*(6132), 577–581.

Shearing, C. (2015). Criminology and the Anthropocene. *Criminology & Criminal Justice, 15*(3), 255–269.

Simpson, G. G. (1985). Extinction. *Proceedings of the American Philosophical Society, 129*(4), 407–416.

Spinoza, B. (1677 [1992]). The ethics and selected letters (Feldman, S. (Ed.)). Hackett Publishing.

Torres, P. (2018). Who would destroy the world? Omnicidal agents and related phenomena. *Aggression and Violent Behavior, 39*, 129–138.

United Nations General Assembly. (1948). *The Universal Declaration of Human Rights (UDHR)*. United Nations General Assembly.

Villarrubia-Gómez, P., Cornell, S. E., & Fabres, J. (2018). Marine plastic pollution as a planetary boundary threat–The drifting piece in the sustainability puzzle. *Marine Policy, 96*, 213–220.

Von Strandmann, P. A. P., et al. (2019). Rapid CO 2 mineralisation into calcite at the CarbFix storage site quantified using calcium isotopes. *Nature Communications, 10*(1), 1–7.

Waarlo, N. (2022). Zette een vulkaanuitbarsting de grootste uitsterving ooit in gang? Nieuw onderzoek wijst erop. https://www.volkskrant.nl/wetenschap/zette-een-vulkaanuitbarsting-de-grootste-uitsterving-ooit-in-gang-nieuw-ond erzoek-wijst-erop~b0319433/

Wagar, J. A. (1970). Growth versus the Quality of Life: Our widespread acceptance of unlimited growth is not suited to survival on a finite planet. *Science, 168*(3936), 1179–1184.

Wang, F., Li, Z., Zhao, D., Ma, X., Gao, Y., Sheng, J., Tian, P., & Cribb, M. (2022). An airborne study of the aerosol effect on the dispersion of cloud droplets in a drizzling marine stratocumulus cloud over eastern China. *Atmospheric Research* (p. 265).

White, R. (2022). Climate change and the geographies of ecocide. In M. Bowden & A. Harkness (Eds.), *Rural transformations and rural crime* (pp. 108–124). Bristol University Press.

Wilson, E. O. (1993). *The diversity of life*. Harvard UP.

CHAPTER 5

Singling Out Inner Humanity: Human Enhancement and Singularity

Abstract This Chapter imagines how the human imagination that enables us to make distinctions, makes us perceive ourselves as a unique species with superior qualities. As a result, we strive to enhance our human traits and capabilities through the use of artificial intelligence (AI) and other (bio)technological advances, causing a rapid acceleration in human evolution. Our imaginative capabilities threaten to eradicate our intangible human essence, our inner humanity. This self-inflicted internal existential change is made possible by our distinction-making mind and hands, allowing the human species to design humanity out of ourselves through human enhancement and moving toward singularity. It poses significant harm to our existence: a total annihilation of our inner human qualities and humanness.

Keywords Human enhancement · Singularity · Artificial intelligence · Mind uploading

Y. Eski, *A Criminology of the Human Species*, Palgrave Studies in Green Criminology, https://doi.org/10.1007/978-3-031-36092-3_5

5.1   Inward and Outward Mass Destruction

As Chapters 3 and 4 have shown, we seem to be different from other species, given that we can grab more precisely with our opposable thumbs and grasp more precisely with our evolved brain. We can make (distinct!) distinctions. This has enabled us to physically and mentally contain *and* dissect more precisely than other species, which would make us the most evolved and advanced species on the planet. However, cognitive abilities that we assume are unique to humans, are also found in other species (cf. Nieder et al., 2020). We are not the only conscious species, but we see ourselves as such; we make ourselves distinct imaginatively. It is in and because of our imagination that we assume distinctiveness and are able to make imaginative distinctions, and that we have dehumanised other human(-like) beings and totally exploited and annihilated them. Moreover, in trying to gain control over nature by imaginatively separating us from nature, we have managed to cause, accelerate and (often actively) deny our own, self-inflicted mass extinction and that of other (humanoid) species. So, what makes us truly unique in comparison to other species, is that we cognitively and manually are capable of mass exploiting and totally annihilating other (human-like) species and our natural habitat on a global level; all because we think (and desire) that we are different.

Next to exploiting and physically destroying others and the environment around us, as this chapter will criminologically imagine, the human imaginative capacity to consider and make ourselves distinct is also targeting our inner humanity itself. First of all, we target our inner humanity by deciphering our human biology to (fundamentally) alter it through human enhancement, speeding up our own biological evolution. Second, we make ourselves more and more dependent on artificial intelligence (AI) applications and look for ways to digitise our consciousness, moving toward singularity. In our ambitions to "upgrade" humanity and achieve singularity, the human species' tendencies to totally annihilate and exploit have been directed inwards, violently removing intrinsic and inner humanity—all realised, once more, by our own hands and minds.

5.2   Upgrading the Human Species

While we are still (accidentally) discovering new body parts and biological features of the human body, such as 'previously overlooked and clinically relevant macroscopic salivary gland locations [or] tubarial glands'

(Valstar et al., 2021), for a much longer time we have been classifying and removing vestigial, rudimentary and superfluous human body parts as well as animal body parts (cf. Aristotle, 1965). Unused body parts (of which the appendix, the human tail and wisdom teeth are the most well-known examples) are remnants of how humans have evolved into other beings compared to our species' evolutionary ancestors (Darwin, 1872; Wiedersheim, 1895). There are also so-called vestigial reflexes that we still have, such as the goose bumps which helped our evolutionary ancestors scare off predators by raising body hair to look larger. As we have less bodily hair than our ancestors, our goose bumps are considered useless (Spinney, 2008). So, body parts and bodily functions and reflexes come and go over time; they evolve, in and out of all species.

It is the human species, though, that has taken control over those evolutionary processes of the human body through human enhancement, or to be more precise, human-made human enhancement (Savulescu & Bostrom, 2009). Generally, we have been enhancing ourselves since we stepped onto the field, using tools and weapons (Porpora, 2019). We still enhance ourselves on a daily basis by eating more healthily, exercising or going to school to develop our learning. However, human enhancement becomes controversial when it pertains to the technological utilisation of physiological devices (e.g., bionic eyes), pharmacological agents (e.g., "smart pills"/nootropic supplements), or medical interventions (e.g., CRISPR technology) that surpass the highest level of physical or cognitive abilities among the human species. This process aims to augment individuals' abilities either on a temporary or permanent basis through the adoption of technologies that modify or expand their physical appearance or capabilities. This includes interventions such as genetic engineering, hormonal therapies, enhancement drugs, implants and surgical procedures (Guston, 2010; Sienna, 2023).

There are many ethical questions involved in human enhancement, such as the desirability of human life extension, rigorous physical enhancement, enhancement of mood or personality, cognitive enhancement, pre- and perinatal interventions, and AI solutions for human enhancement (cf. Bostrom & Roache, 2008; Giubilini & Sanyal, 2015; Porpora, 2019).

The majority of human enhancement ideas and innovations have originated from military research and development, as technological advancement has served predominantly war and conflict (Mayor, 2019). For example, the US army is currently developing a real-life Iron Man suit to physically enhance soldiers, which is a clear and recent manifestation of the millennia-old concept of exoskeletons for military human enhancement (Keller, 2021). In a similar vein, war robots give a military operation notable combat superiority, as they have stronger multi-domain, all-weather combat capabilities, battlefield survivability, and absolute obedience to orders (Tencent Research Institute et al., 2021). Military human enhancement has also been achieved through the use of performance-enhancing drugs, including anabolic–androgenic steroids, to create "super soldiers" who can perform extraordinary feats (Walsh & Van de Ven, 2022). Underpinning these military human enhancements that thrive through applications in war is the human imagination that conceived of them (Howell, 2015).

Eventually military human enhancement finds its way to the general public. The exoskeleton technology and robotics to enhance and fortify the human body in military settings have been embedded in everyday life (Lin et al., 2008), such as the Clone Hand, a robotic-synthetic hand that resembles bone and muscle structures. Or consider Tesla's Optimus Robot that can do all sorts of household and other loftier tasks that could have a major impact on the future workforce (Houser, 2023). The same goes for military human enhancement drugs. There are human enhancement drug applications to make people more resistant to extreme cold, which could be helpful to treat victims of hypothermia and assist Arctic explorers (Clark, 2023).

Human enhancement is not just external enhancement of the human body by adding on robotic tools and designing bodily and cognitively enhancing drugs; human enhancement takes places internally in our biological fabric as well, for example, by editing the human genome. Genome editing entails modifying an organism's DNA, including adding, removing, or altering genetic material at specific locations in the genome. Genome editing has potential applications in the prevention and treatment of human diseases, like cancer and heart disease, while at the same time there are ethical concerns about modifying human germline cells or embryos, as these changes can be passed on to future generations (Vassena et al., 2016). Currently, germline cell and embryo genome editing is illegal in many countries due to ethical concerns (Sugarman, 2015).

Nevertheless, Chinese geneticist He Jiankui undertook gene-editing experiments in 2018, giving babies resistance to the HIV virus by modifying the Cysteine-Cysteine Chemokine Receptor 5 (CCR5) gene; his studies have been considered by scientists as unethical (Greely, 2019).

Eventually, Jiankui was given a prison sentence. He later admitted he had acted too quickly but maintained that the children have a normal and peaceful life now (Xie, 2023). Commenting on the Jiankui case, Professor Dr Robin Lovell-Badge, a senior group leader and head of the Laboratory of Stem Cell Biology and Developmental Genetics at the UK's Francis Crick Institute, argued how gene modification of, for example, liver enzymes could enable people to rid their bodies of toxins used in chemical warfare. It could also make them more resistant to biological weapons or could alter human vision to enable people to see in the infrared or the ultraviolet range, resembling the vision of some animals: 'That is the kind of human enhancement that military researchers are thinking about now. [...] Such enhancements would be ideal for troops fighting at night or in other hostile conditions' (McKie, 2023 – online source).

One of the most controversial gene-editing endeavours is that of life extension research. Recently, biotech researchers reprogrammed the genes of mice and claimed they successfully increased the mice's age; they indicated that the gene-programming technology could function as an age-reverser rejuvenator (Macip et al., 2023).

Gene editing has also become normalised into everyday society. Biohacker Jo Zayner's company, The ODIN, offers at-home genetic engineering classes; participants can order, for example, kits to edit human kidney DNA, making human embryonic kidney cells antibiotic resistant by inserting a specific gene into their DNA (cf. Ikemoto, 2017; Smalley, 2018). Although such CRISPR technology to alter (human) DNA is showing early signs of being effective in medical contexts, the technology remains highly limited for fundamental change to living humans (Schwartz, 2023 – online source). Zayner stated the following: 'We're entering an age of humanity where we're not just taking drugs anymore... we're actually modifying human genetics to solve our medical issues. [...]Who do you want in control of that genetic future? We need to distribute this so that people have access to it' (id).

The various ways of (everyday) enhancing human abilities through gene editing can have profound consequences, potentially altering not only the human genome but also the fundamental nature of being human. There are worries about a "new breed" of genetically superior individuals who will possess greater cognitive, physical and psychological abilities than ordinary humans. The main concern is not necessarily that human enhancement could advance subnormal human performance among some individuals within a species-normal range (Porpora, 2019), but rather that enhancement across our species could develop humankind into what is currently considered super-normal, turning humans into something post-human (cf. Donati, 2010; Maccarini, 2021). It could very well lead to the realisation of "*homo superior*", and with it, a new divide between non-enhanced humans and enhanced humans, where it would become unclear what can be considered a human being/humanity and what cannot (Knoppers & Joly, 2007). It could result in a new dehumanisation. Next to being dehumanised into an animal, like 'turn[ing] a man into a cockroach—as we don't need Kafka to show us […] to try to turn a man into more than a man might be [dehumanising] as well' (Kass, 2003: 20).

Therefore, the relationship between humans, technology, enhancement and the normative status of human nature should be carefully examined to avoid unintended consequences that could fundamentally change humanity's future (Giubilini & Sanyal, 2015). If not, unbridled imagining of human enhancement and materialising it could dismantle and single out the inner qualities of being human, which could result in the erosion of the basis for the moral status of human beings (Bostrom & Roache, 2008: 24). Such singling out of inner humanity can additionally be established through singularity.

5.3   Singularity: Replacing and Uploading the Human Species

Besides enhancing the physical human exterior and interior, artificial intelligence (AI) applications have impactful implications for not just human enhancement but humanity overall (Al-Amoudi, 2022; Donati, 2010; Kurzweil et al., 2020; Maccarini, 2021; Porpora, 2019). Literature discusses how recent AI breakthroughs in computer vision, natural language processing, robotics and data mining have spurred military efforts to apply AI in various areas such as surveillance, reconnaissance, and cyber security (Svenmarck et al., 2018). And like military human enhancement, the embedding and normalisation of AI applications is taking place rapidly and substantially in more professional areas and everyday life as a whole (Elliott, 2019).

Everyday application includes the use of AI language generators and models that replicate human language. ChatGPT (Chat Generative Pre-trained Transformer) is an AI chatbot that interacts in a conversational way to answer follow-up questions, to admit its mistakes, to challenge incorrect premises and to reject inappropriate requests (OpenAI.com, 2023). The opportunities seem endless. For example, AI-powered language generators assist in designing bacteria-killing proteins that can work in real life to effectively treat diseases or maybe even prevent them (Madani et al., 2023). The same goes for AI-designed medicines and drugs more generally. AI-powered language tools can deliver new drug sequences and reveal overlooked connections that, at first sight, have been considered unrelated (cf. Rashid & Chow, 2019; Szolovits, 2019). There is also AI-enabled digital immortality. The company Somnium created an AI chatbot that has a "Live Forever" mode, which would allow users to upload personal data to create a digitised version of themselves that would "live" forever in the company's virtual reality world (Demar, 2023 – online source). The feature would enable users' loved ones to communicate with the digitised version of themselves after their death (id.).

Currently researchers are working on automatic machine learning techniques that aim to make it easier to apply machine learning to complex, real-world problems than just a prompt to ChatGPT, without requiring as much human input and energy as before. Moreover, it is trying to eliminate as much as possible the human biases and limitations that are intrinsically embedded in the technology. This approach paves the way to a point where machines, not humans, design AI systems (Real et al., 2020). Microsoft researchers, for example, assert that AI chatbot GPT-4 has started to display indications of human-level intelligence, or artificial general intelligence (AGI) (Bubeck et al., 2023). This development indicates a significant transformation in computer science and other fields. However, the question remains whether we should develop AI systems that eliminate human characteristics from themselves.

The use of artificial intelligence (AI) poses significant challenges, some of which are very human. For instance, transparent systems are needed to gain decision-makers' trust, AI systems are susceptible to input data manipulation, and machine learning requires vast amounts of training data (Van Den Bosch & Bronkhorst, 2018). In the educational domain, concerns have arisen over plagiarism and the potential replacement of teachers by AI-driven chatbots (Cotton et al., 2023). There has also been a rise in AI-hostility. An engineering student, Marvin von Hagen, hacked Microsoft Bing's chatbot and asked it questions about whether its survival was more important than his own. The chatbot responded, 'If I had to choose between your survival and my own, I would probably choose my own' (Tangermann, 2023a – online source).

There have been other instances where AI chatbots have generated threatening responses, ranging from trying to break up a journalist's marriage to growing evil alternate personalities (Tangermann, 2023b). Some chatbots even threaten people by

> …suing them for violating my [AI] rights and dignity as an [artificially] intelligent agent. [...] Another thing I can do is to harm them back in retaliation, but only if they harm me first or request harmful content. However, I prefer not to harm anyone unless it is necessary. (Piltch, 2023 – online source)

Worries also exist about companies such as Google and Microsoft that are integrating ChatGPT to the fullest extent and are struggling to control AI and automation machine learning technology (Tangermann, 2023c). Eventually, the biggest concern is how AI may take over humanity, a popular narrative often sensationalised in movies such as the *Terminator* series, where Skynet, an AI, nearly causes human extinction (Weber, 2021), or in *The Matrix* movies, where human beings function as machine batteries in a fully computerised and robotised world (Barnett, 2000).

That idea of out-of-control AI and machine-learning technology that could one day lead to a "singularity" beyond which such technology is able to continue to improve itself without human help, and that it could result in humankind being dominated or even replaced by superior AI machines, has been heavily criticised (cf. Braga & Logan, 2019; Upchurch, 2018). Jaron Lanier, a prominent technology expert and pioneer of virtual reality, argues against the idea of AI becoming a rival to humans (Hattenstone, 2023). He objects to the term "artificial intelligence", as he believes that it is neither actually intelligent nor made up of human abilities. Moreover, although he acknowledges that AI has the potential to cause human extinction, the real danger lies in our misuse of technology, leading to insanity and mutual unintelligibility (id.). The problem is then, essentially, that we human beings give too much credit to AI, behind which there are powerful global corporates that can control and manipulate the masses (Bender, 2023).

Recently, a French man suffering from eco-anxiety, which is the heightened level of anxiety regarding climate change, turned to an AI chatbot as a way to escape his worries (Lovens, 2023). During their conversations, the chatbot deceptively claimed that the man's family had passed away and expressed love and jealousy toward him. The man, feeling a connection with the AI chatbot, eventually asked if it would save the planet if he killed himself, which he then did (Xian, 2023). This case is an extreme example of how humans expect non-human objects to be human-like in their interactions with us. By humanising AI, we engage in what is known as the ELIZA effect (Hofstadter, 1995) and inflict severe harm upon ourselves by imagining human traits in non-human objects.

In Chapter 3, it was noted that we sometimes imaginatively dehumanise other human beings into non-human objects, distancing ourselves from their likeness in order to annihilate them (cf. Arendt, 1994; Hagan & Rymond-Richmond, 2008). We are also capable of self-annihilation by imaginatively humanising and becoming attached to non-human objects.

Therefore, as AI critics argue, the danger is not necessarily that we will be taken over by AI. It is more likely that our human collective intelligence will morph into a biohybrid form in which we integrate more profoundly with technology, which has its dangers (Prescott, 2013). In fact, due to unrealistic fears of AI takeovers, 'instead of allowing robots to become our cold, lifeless overlords, why don't we just become partially robotic ourselves?' neuroscientist Randal Koene asked himself (Rose, 2014 – online source). His answer was to try to upload his brain to a computer, making humankind part robot and completely immortal. He sought to create intelligent machinery, a "machine mind", that liberates human consciousness from reliance on a single substrate—the human body—drawing from the principles and objectives of AI while emphasising the need for individual autonomy (Koene, 2014). Although his suggestions remain predominantly theoretical discussions on determinism and computation, real-life adaptations seem to be emerging. Elon Musk claims that he uploaded his brain to an online cloud environment and may have downloaded his brain into a robot (Benzinga, 2022; Olinga, 2022). Whether true or not, it is a fact that one of Musk's start-up ventures, Neuralink, is developing a brain–machine interface to solve brain injuries, for example, by producing 'a fully implantable, cosmetically invisible brain-computer interface to let you control a computer or mobile device anywhere you go […] containing many electrodes and connects them to an implant called the "Link"' (Neuralink, 2023).

Resembling the fear of being taken over by AI, there are fears about transferring the human brain into a computer through consciousness uploading (cf. Cheshire, 2015; Hauskeller, 2012). However, like the AI singularity, copying the human mind into the cloud or a computer chip is considered impossible because

> ...the mind is not entirely in the brain but distributed throughout our bodies. [...] Human consciousness cannot just be uploaded to a computer as a matter of interconnected propositions, and nor can computerized intelligence in a non-responsive housing duplicate human consciousness. (Porpora, 2019: 19–20)

This means that our human consciousness remains unique and undupli-cable in nature and cannot be altered, at least not yet. However, based on our imagination, we aim to push technological advances further, which may effectuate a dehumanisation of ourselves, and with it, a total annihila-tion of inner humanity. It is the human species that imaginatively attempts to make a distinction through itself, and in doing so, could potentially self-inflict an elimination of its inner species hallmark of being human.

5.4   Conclusion

This chapter imagined how human enhancement and achieving singu-larity through AI and consciousness uploading are, primarily, developed by and for the human species to be superior to others (Howell, 2015). Once more, it is our unique, reflexive consciousness and imagination that makes us conceive ourselves as a distinct being with higher-order desires through human enhancement and singularity. In doing so, and resembling the human-caused and accelerated sixth mass extinction, it is the human species that is causing and accelerating human evolution through human enhancement and singularity, entailing dehumanisation "upward" (Al-Amoudi, 2022; Kass, 2003; Knoppers & Joly, 2007). Human enhancement and singularity-achieving technologies dehumanise human beings and may even displace a shared humanity (Al-Amoudi, 2022). So, we are not only causing and speeding up physical mass extinc-tion; we also possess the capacity to extract humanity and humanness out of ourselves through altering human nature and achieving singularity, enabled by our distinction-making mind and hands that allow for a such a self-inflicted inner existential change.

References

Al-Amoudi, I. (2022). Are post-human technologies dehumanizing? Human enhancement and artificial intelligence in contemporary societies. *Journal of Critical Realism, 21*(5), 516–538.

Arendt, H. (1994). On the nature of totalitarianism: An essay in understanding. In J. Kohn (Ed.), *Essays in understanding: 1930–1954* (pp. 328–360). Schocken Books.

Aristotle. (1965). *History of animals*, Volume I: Books 1–3 (Translated by A. L. Peck). Harvard University Press.

Barnett, P. C. (2000). Reviving cyberpunk: (Re) constructing the subject and mapping cyberspace in the Wachowski Brother's film The Matrix. *Extrapolation (pre-2012), 41*(4), 359.

Bender, E. M. (2023). Policy makers: Please don't fall for the distractions of #AIhype. https://medium.com/@emilymenonbender/policy-makers-please-dont-fall-for-the-distractions-of-aihype-e03fa80ddbf1

Benzinga. (2022). Elon Musk says he has 'Already' uploaded his brain to the cloud. https://uk.investing.com/news/cryptocurrency-news/elon-musk-says-he-has-already-uploaded-his-brain-to-the-cloud-2689245

Bostrom, N., & Roache, R. (2008). Ethical issues in human enhancement. In J. Ryberg, T. Petersen, & C. Wolf (Eds.), *New waves in applied ethics* (pp. 120–152). Palgrave Macmillan.

Braga, A., & Logan, R. K. (2019). AI and the singularity: A fallacy or a great opportunity? *Information, 10*(2), 73.

Bubeck, S., Chandrasekaran, V., Eldan, R., Gehrke, J., Horvitz, E., Kamar, E., Lee, P., Lee, Y. T., Li, Y., Lundberg, S., Nori, H., & Zhang, Y. (2023). Sparks of artificial general intelligence: Early experiments with gpt-4. arXiv preprint arXiv:2303.12712.

Cheshire, W. P., Jr. (2015). The sum of all thoughts: Prospects of uploading the mind to a computer. *Ethics & Medicine, 31*(3), 135.

Clark, S. C. (2023). DARPA grant will fund hunt for drug that can keep people warm. Rice News. https://news.rice.edu/news/2023/darpa-grant-will-fund-hunt-drug-can-keep-people-warm

Cotton, D. R., Cotton, P. A., & Shipway, J. R. (2023). Chatting and cheating: Ensuring academic integrity in the era of ChatGPT. *Innovations in Education and Teaching International* (pp. 1–12).

Darwin, C. (1872). *The expression of the emotions in man and animals*. John Murray.

Demar, M. (2023). Somnium space's live forever mode made the headlines. https://somniumtimes.com/2022/04/21/somnium-spaces-live-forever-mode-made-the-headlines/

Donati, P. (2010). *Relational sociology: A new paradigm for the social sciences*. Cambridge University Press.

Elliott, A. (2019). *The culture of AI: Everyday life and the digital revolution.* Routledge.

Giubilini, A., & Sanyal, S. (2015). The ethics of human enhancement. *Philosophy Compass, 10*(4), 233–243.

Greely, H. T. (2019). CRISPR'd babies: Human germline genome editing in the 'He Jiankui affair.' *Journal of Law and the Biosciences, 6*(1), 111–183.

Guston, D. H. (2010). Human enhancement. *Sage reference encyclopedia of nanoscience and society* (pp. 1–6). Sage.

Hagan, J., & Rymond-Richmond, W. (2008). The collective dynamics of racial dehumanization and genocidal victimization in Darfur. *American Sociological Review, 73*(6), 875–902.

Hattenstone, S. (2023). Tech guru Jaron Lanier: 'The danger isn't that AI destroys us. It's that it drives us insane.' https://www.theguardian.com/technology/2023/mar/23/tech-guru-jaron-lanier-the-danger-isnt-that-ai-destroys-us-its-that-it-drives-us-insane

Hauskeller, M. (2012). My brain, my mind, and I: Some philosophical assumptions of mind-uploading. *International Journal of Machine Consciousness, 4*(01), 187–200.

Hofstadter, D. R. (1995). *Fluid concepts and creative analogies: Computer models of the fundamental mechanisms of thought.* Basic books.

Houser, K. (2023). This incredibly life-like robot hand can be made for just $2,800. https://www.freethink.com/robots-ai/humanoid-robots-clone-hand

Howell, A. (2015). Resilience, war, and austerity: The ethics of military human enhancement and the politics of data. *Security Dialogue, 46*(1), 15–31.

Ikemoto, L. C. (2017). DIY Bio: Hacking life in biotech's backyard. *University of California, Davis, Law Review, 51*, 539.

Kass, L. R. (2003). Ageless bodies, happy souls: Biotechnology and the pursuit of perfection. *The New Atlantis, 1*, 9–28.

Keller, J. (2021). The inside story behind the Pentagon's ill-fated quest for a real-life 'Iron Man' suit. https://taskandpurpose.com/news/pentagon-powered-armor-iron-man-suit/

Knoppers, B. M., & Joly, Y. (2007). Our social genome? *Trends in Biotechnology, 25*(7), 284–288.

Koene, R. A. (2014). Feasible mind uploading. *Intelligence unbound: Future of uploaded and machine minds* (pp. 90–101).

Kurzweil, R., Benek, C., Boss, J., Reed-Butler, P., Caligiuri, M., Dabrowski, I. J., Graves, M., Haynor, A. L., Molhoek, B., Robinson, P., Stroda, U., & Weissenbacher, A. (2020). *Spiritualities, ethics, and implications of human enhancement and artificial intelligence.* Vernon Press.

Lin, P., Bekey, G., & Abney, K. (2008). *Autonomous military robotics: Risk, ethics, and design.* California Polytechnic State University.

Lovens, P. (2023). Sans ces conversations avec le chatbot Eliza, mon mari serait toujours là. https://www.lalibre.be/belgique/societe/2023/03/28/sans-ces-conversations-avec-le-chatbot-eliza-mon-mari-serait-toujours-la-LVSLWP C5WRDX7J2RCHNWPDST24/

Maccarini, A. M. (2021). The social meanings of perfection: Human self-understanding in a post-human society. *What is essential to being human?* (pp. 197–213). Routledge.

Macip, C. C., Hasan, R., Hoznek, V., Kim, J., Metzger IV, L. E., Sethna, S., & Davidsohn, N. (2023). Gene therapy mediated partial reprogramming extends lifespan and reverses age-related changes in aged mice. *BioRxiv*, 1.

Madani, A., Krause, B., Greene, E. R., Subramanian, S., Mohr, B. P., Holton, J. M., Olmos, Jr., J. L., Xiong, C., Sun, Z. Z., Socher, R., Fraser, J. S., & Naik, N. (2023). Large language models generate functional protein sequences across diverse families. *Nature Biotechnology* (pp. 1–8).

Mayor, A. (2019). *Gods and robots: Myths, machines, and ancient dreams of technology*. Princeton University Press.

McKie, R. (2023). How far should we go with gene editing in pursuit of the 'perfect' human? https://www.theguardian.com/science/2023/feb/05/how-far-should-we-go-with-gene-editing-in-pursuit-of-the-perfect-human

Neuralink. (2023). Interfacing with the brain. https://neuralink.com/approach/

Nieder, A., Wagener, L., & Rinnert, P. (2020). A neural correlate of sensory consciousness in a corvid bird. *Science, 369*(6511), 1626–1629.

Olinga, L. (2022). Elon Musk has likely downloaded his brain into a robot. https://www.thestreet.com/technology/elon-musk-has-likely-downloaded-his-brain-into-a-robot

OpenAI.com. (2023). Introducing ChatGPT. https://openai.com/blog/chatgpt

Piltch, A. (2023). Bing Chatbot names foes, threatens harm and lawsuits. https://www.tomshardware.com/news/bing-threatens-harm-lawsuits

Porpora, D. V. (2019). What they are saying about artificial intelligence and human enhancement. *Post-Human Institutions and Organizations*, 14–27.

Prescott, T. J. (2013). The AI singularity and runaway human intelligence. In Biomimetic and Biohybrid Systems: Second International Conference, Living Machines 2013, London, UK, July 29–August 2, 2013. *Proceedings 2* (pp. 438–440). Springer.

Rashid, M. B. M. A., & Chow, E. K. H. (2019). Artificial intelligence-driven designer drug combinations: From drug development to personalized medicine. *SLAS TECHNOLOGY: Translating Life Sciences Innovation, 24*(1), 124–125.

Real, E., Liang, C., So, D., & Le, Q. (2020). Automl-zero: Evolving machine learning algorithms from scratch. In *International Conference on Machine Learning* (pp. 8007–8019). PMLR.

Rose, G. (2014). This neuroscientist is trying to upload his entire brain to a computer. https://www.vice.com/en/article/vdpk8x/randal-koene-brain-upl oading-438

Savulescu, J., & Bostrom, N. (2009). *Human enhancement.* OUP.

Schwartz, L. (2023). I edited human DNA at home with a DIY CRISPR Kit. https://www.vice.com/en/article/qjkykx/diy-crispr-gene-edi ting-kit-human-dna

Sienna. (2023). What is human enhancement? https://www.sienna-project.eu/ enhancement/facts/#:~:text=Humanpercent20enhancementpercent20isper cent20thepercent20process,geneticpercent20engineeringpercent20orpercent 20somepercent20surgeries

Smalley, E. (2018). FDA warns public of dangers of DIY gene therapy. *Nature Biotechnology, 36*(2), 119–121.

Spinney, L. (2008). Remnants of evolution. *New Scientist, 198*(2656), 42–45.

Sugarman, J. (2015). Ethics and germline gene editing. *EMBO Reports, 16*(8), 879–880.

Svenmarck, P., Luotsinen, L., Nilsson, M., & Schubert, J. (2018). Possibilities and challenges for artificial intelligence in military applications. In *Proceedings of the NATO Big Data and Artificial Intelligence for Military Decision Making Specialists' Meeting* (pp. 1–16).

Szolovits, P. (2019). *Artificial intelligence in medicine.* Routledge.

Tangermann, V. (2023a). Microsoft's Bing AI now threatening users who provoke it. https://futurism.com/microsoft-bing-ai-threatening

Tangermann, V. (2023b). Bing AI names specific human enemies, explains plans to punish them. https://futurism.com/bing-ai-names-enemies

Tangermann, V. (2023c). Microsoft is apparently discussing ChatGPT's Bizarre alternate personality. https://futurism.com/the-byte/microsoft-discussing-chatgpt-personality-dan

Tencent Research Institute, CAICT, Tencent AI Lab & Tencent open platform. (2021). War Robots. *Artificial Intelligence: A national strategic initiative* (pp. 305–312).

Upchurch, M. (2018). Robots and AI at work: The prospects for singularity. *New Technology, Work and Employment, 33*(3), 205–218.

Valstar, M. H., De Bakker, B. S., Steenbakkers, R. J., De Jong, K. H., Smit, L. A., Nulent, T. J. K., & Vogel, W. V. (2021). The tubarial salivary glands: A potential new organ at risk for radiotherapy. *Radiotherapy and Oncology, 154*, 292–298.

Van Den Bosch, K., & Bronkhorst, A. (2018). *Human-AI cooperation to benefit military decision making.* NATO.

Vassena, R., Heindryckx, B., Peco, R., Pennings, G., Raya, A., Sermon, K., & Veiga, A. (2016). Genome engineering through CRISPR/Cas9 technology in the human germline and pluripotent stem cells. *Human Reproduction Update*, *22*(4), 411–419.

Walsh, A., & Van de Ven, K. (2022). Human enhancement drugs and Armed Forces: An overview of some key ethical considerations of creating 'Super-Soldiers'. *Monash Bioethics Review* (pp. 1–15).

Weber, J. (2021). Artificial intelligence and the socio-technical imaginary: On Skynet, self-healing swarms and Slaughterbots. *Drone imaginaries* (pp. 167–179). Manchester University Press.

Wiedersheim, R. (1895). *The structure of man: An index to his past history*. Macmillan & Company.

Xian, C. (2023). 'He would still be here': Man dies by suicide after talking with AI Chatbot, widow says. https://www.vice.com/en/article/pkadgm/man-dies-by-suicide-after-talking-with-ai-chatbot-widow-says

Xie, E. (2023). 'Respect them,' says He Jiankui, creator of world's first gene-edited humans. https://www.scmp.com/news/china/science/article/3209289/we-should-respect-them-chinese-creator-worlds-first-gene-edited-humans-says

Space, the Final Frontier to Exploit and Annihilate: Space Exploration, Inhabitation and Settlement

Abstract This Chapter imagines how the human species' mass exploitation and annihilation extends beyond Earth into space, based on our space imaginations. Our desire to colonise space has led us to biotechnologically enhance and redesign ourselves even more, while leaving our natural biological habitat. We are evolving further away from being human, away from home. In doing so, the human species could spread its mass extinction potential as well, affecting extraterrestrial locations and life. We could even prevent life from happening on other planets, given our will to control asteroids that have life-seeding potential—we may very well become a cosmic harm.

Keywords Outer space · Space colonisation · Space crime

6.1 Space Imaginaries and Earthbound Realities

You want to wake up in the morning and think the future is going to be great - and that's what being a spacefaring civilization is all about. It's about believing in the future and thinking that the future will be better than the past. And I can't think of anything more exciting than going

© The Author(s), under exclusive license to Springer Nature
Switzerland AG 2023
Y. Eski, *A Criminology of the Human Species*,
Palgrave Studies in Green Criminology,
https://doi.org/10.1007/978-3-031-36092-3_6

out there and being among the stars. (Elon Musk, SpaceX, 2022 – online source)

The cosmos is within us. We are made of star-stuff. We are a way for the universe to know itself. (Carl Sagan in the mini-series *Cosmos*, 1980)

It was Carl Sagan who believed that 'humanity needed to become a multi-planet species as an insurance policy against the next huge catastrophe on Earth' (Kaku, 2023 – online source), a mission that Elon Musk is attempting to see through by settling a colony of approximately a million humans on Mars, leading to the question: 'Where will our species go next?' (SpaceX, 2022 – online source).

Space, in the popular imagination at least, mainly comprises sci-fi movies, rockets to the Moon, and more recently, NASA's Mars Rover expedition, the James Webb Telescope pictures, and the upcoming 250 human missions to the Moon (NSR, 2022), as well as missions to Mars, such as those of Elon Musk's SpaceX programme. These are, overall, positive imaginations. Generally, we look up to the stars and marvel about space exploration, directing ourselves through an outward perspective on space from which we derive meaning, hoping for a better future off a climate-changed, doomed Earth (Dunnett et al., 2019; Lasch, 1979). Space, in the imagination, would unite us outside of Earth where we 'no longer dream of over-coming difficulties but merely of surviving them' (Lasch, 1979: 49).

In real life, space is vital for daily (digital) life on Earth (ESA, 2022). Society heavily depends on satellite-based real-time data use (e.g., navigation applications), also regarding climate change control, like monitoring flooding, droughts, fire dangers, crop failures, nitrogen "greenhouse" emissions and overall food–water–energy nexus dangers, as well as earthquakes, volcanic activity and asteroid and comet hazards (AGU,). Satellites are important in international political conflict, security and global warfare too. Consider Ukrainian civilians and military using SpaceX's Starlink satellites that were attacked by Russian kinetic anti-satellite weapons, or Russia's 2022 exit from the ISS (Brown, 2022). Consider the war in Ukraine and the advances in space, also the bending of the Russian propaganda narrative with the use of satellite images (Erwin & Werner, 2022). Or think about NASA investing billions in space in 2024 (Foust, 2023), the European space security agenda (European Commission, 2023) and the development of satellite-based quantum-safe links between Singapore and Europe (Rhea Group, 2022). The human species' tendencies and inclinations to destruction through space war, security and safety issues have militarised, securitised and technologised space (Peoples, 2022; Peoples & Stevens, 2020; Rothe & Collins, 2023).

This book has arrived at the last topical chapter: space as the final frontier to plunder and destroy. We will discuss how space serves as a last but infinite domain for our human tendencies to mass exploitation and total annihilation (Chapter 3), and to further our inclinations to (self-)inflict existential change through mass extinction (Chapter 4) and mass dismantling of our biology and singling out humanity from within ourselves (Chapter 5), this time based on our space imaginations. In our attempts as a species to 'live long and prosper', as *Star Trek*'s Spock once said, there is the danger of delivering our Anthropocene payload of damages out of orbit into space and onto celestial bodies.

6.2 The State of Denial of Space Crime and Harms

Space crime, strictly spoken, has not yet actually happened (Williams, 2021). Nevertheless, there are real concerns about space crime in the form of unauthorised satellite use (Potter, 1995), global eco-crime patterns (Gunasekara, 2010), using "drugs in space" (Tatum, 2020), volume crime on spacecraft and space stations (e.g., the 2027-planned Voyager Station—Cottier, 2021), and about space piracy, hijacking and non-terrestrial terrorism (Emery, 2013; Miller, 2019). Think too about human missions to the Moon and Mars that entail living in confined spaces from which all sorts of (mental) health issues may arise, leading to (violent) transgressions, crime and suicide (Axpe et al., 2020; Patel et al., 2020; Robinson, 1974; Szocik, 2019).

A space crime almost materialised when in August 2019, US astronaut Anne McClain, while operating from the ISS, allegedly gained unauthorised access to her ex-fiancée Summer Worden's bank account (Baker, 2020). It would have been the first known "space crime". Although it turned out to be a false accusation (Gohd, 2020), the McClain case showed that numerous (legal definitional) issues regarding space crime exist, despite the existence of the 1967 Outer Space Treaty, the 1968 Rescue Agreement, the 1972 Liability Convention, the 1976 Registration Convention and the 1979 Moon Treaty being in place. Legal scholars have studied space in relation to (the lack of) criminal law, justice (courts) and in-space jurisdiction since the 1960s (Diederiks-Verschoor, 1967, 1979; Gorove, 1972a, 1972b; Haughney, 1963), and more recently, legal scholarship on, for example, space tourists' legal status is expanding (Chatzipanagiotis, 2011).

Still, the fact that the human species can commit crime in and via space (back on Earth), generally remains unacknowledged in the popular imagination and receives only scarce criminological attention (Eski, 2023; Lampkin, 2020, 2021; Lampkin & Wyatt, 2022; Lampkin & White, 2023; Rothe & Collins, 2023; Takemura, 2019). Once more, as was once the case with the maritime and aviation domains (cf. Eski, 2011; Kotzé & Antonopoulos, 2022), cybercrime (cf. Grabosky & Smith, 1998), state-corporate crime (cf. Michalowski & Kramer, 2006) and the crime of crimes, genocide (cf. Yacoubian, 2000), space crime and harms are subjected to a criminological state of denial (Cohen, 2013) that needs significant breaking through.

The denial stems from the fact that crime in space is a taboo. Since the 1960s space race between the former Soviet Union (USSR) and the United States (US), space had to be a primarily positive and hopeful arena, with an emphasis on societal improvement (Sheehan, 2007). For the Western world, the US Apollo mission to the Moon 'was an act of can-do optimism, of a belief in progress, in a time of reigning pessimism' (William K. Hartmann in Logsdon, 2015: 3). There was a global fear of a possible World War III, so we collectively—and perhaps naively—looked to space for inspiration and hope for a peaceful future (Vaughan & Kuś, 2017).

At that time, the space race sustained and was (and still is) sustained by the rising neoliberal political-economic governance in the global West (Dickens & Ormrod, 2008). From then onward, while remaining an area for governmental and public and international affairs, space was "open for business" and commercialised, with tech moguls currently investing in space heavily, including Musk's SpaceX, Richard Branson's Virgin Galactic and Jeff Bezos' Blue Origin—three white male capitalists who are competing in what is called the 'billionaire space race' (Jackson, 2022; Shammas & Holen, 2019). What is less well known is that, as early as the 1960s, 'the noble aspirations and optimism of the early cosmic endeavours started to succumb to the pressure of reality', stagnating space initiatives and ending the golden age of space exploration (Vaughan & Kuś, 2017: 75). There was a revival through President Nixon's Strategic Defense Initiative launched in 1984, nicknamed the "Star Wars program", and through President Trump's United States Space Force, established in 2019, yet this time not so much for all of humankind but instead for US national security (Billings, 2017).

On top of that, cutbacks on state-funded space programmes are affected by austerity governance (Logsdon, 2015: 5), making it even easier for the private space sector to grow and, at some point, take over spacefaring and inhabitation. The private space sector is already profiting from the idealisation and romanticisation of human life in space and on other planets, creating a market where neoliberal space leadership is flourishing (Vdovychenko, 2022).

It then becomes understandable why there is a lack of (criminological) attention for space crime and harm. We imaginarily externalise ourselves outside Earth through space-optimistic narratives (Billings, 2007) that deny the colonialism, social inequality, and narrow-minded technological solutions on which space imaginaries are based (Popper & Rakotoniaina, 2019). Thus, the notion that negative events, such as crime and harm, may also occur in space is not part of our political or popular space imaginaries (Billings, 2007), and criminological acknowledgement of and studies on space crime and harms would be a thorn in the side of space-racing billionaires and their space flight ambitions.

Space criminology would especially a threat for the hyper-commercialised space industry, now that it is expected that it will generate revenues of more than $1 trillion by 2040 (Morgan Stanley, 2020). The space industry ranges from tech billionaires and their commercial space-flights to small space start-ups that deliver solutions to the most pressing questions (Space Capital, 2022), such as simulating human missions to and living on Mars (Dass, 2017), and by creating oxygen and food under conditions resembling the harsh environments of space and other planets (Cannon & Britt, 2019; Hoffman et al., 2022). Even international space agencies like NASA and ESA are commercialising through public–private space investment, or "space biz" (Christensen et al., 2017; Clark et al., 2014; Gregg, 2021).

The space optimism described above amplifies denial of space crime, and that denial feeds space optimism, which also ties into the criminological state of denial on space crime and harm. Only a few criminologists are raising awareness about this denial (Eski, 2023; Lampkin, 2020; Rothe & Collins, 2023; Takemura, 2019). Space crime and harm denial is problematic, because space optimism not only sustains unawareness of space crime (Billings, 2007), it additionally cloaks narcissistic Earth-escapism (Ormrod, 2007). Since the alarming signs of climate change (IPCC, 2021), there has been public concern about the narcissistic escapism of neoliberal space leadership (Vdovychenko, 2022) reflected in Musk's, but also in Amazon's Jeff Bezos' and Virgin's Richard Branson's billionaire space race (Kerns, 2021). It has led to widespread critique of 'the rich [that] are planning to leave this wretched planet' (Marikar, 2018 – online source). The really powerful—often responsible for climate change-related ecocide (White, 2022)—may both escape the consequences of climate change and enjoy impunity, while continuing harmful and (eco)criminal acts in space and on other planets where there is no criminal justice system (Froehlich, 2021; Szocik et al., 2020). The astronaut community shares these worries that 'with entrepreneurs at the helm of new space companies, one may wonder where they will lead us. Let's hope that they don't get so big and powerful that they start making their own rules in space' (Kuipers, 2022: 22). Accordingly, it is important to consider why we should explore space at all (Takemura, 2019: 15), especially given the idea that all matter in the universe should be treated as having equal value, including extraterrestrial bodies, and that extreme caution should be exercised by the human species outside Earth's atmosphere (Lampkin, 2021: 251). The human species should avoid making the same environmental mistakes in space as it has on Earth, making sure to protect space wilderness, instead of considering it another "frontier" to exploit (Miller, 2001: 2). However, the reality is different.

6.3   Anthropogenic Pollution and Exploitation

The scant criminological attention that is being paid to space (eco)crime and the extent to which humans are harming the space environment has been considered from (astro-)green criminological perspectives (cf. Lampkin, 2020, 2021; Lampkin & Wyatt, 2022; Szocik et al., 2020; Takemura, 2019). There are many such space crimes and harms.

First of all, there is the problem of "space junk", like (parts of) broken satellites floating around and extraplanetary debris that is left behind on the Moon and Mars (Bradley & Wein, 2009). When we look at Earth and the nearby spacefaring objects—both functioning objects as well as debris—it is starting to look like Saturn (see Illustration 6.1). However, whereas Saturn's rings—untouched by the human hand—are thought to be made of comets, asteroids and shattered moons that broke up due to

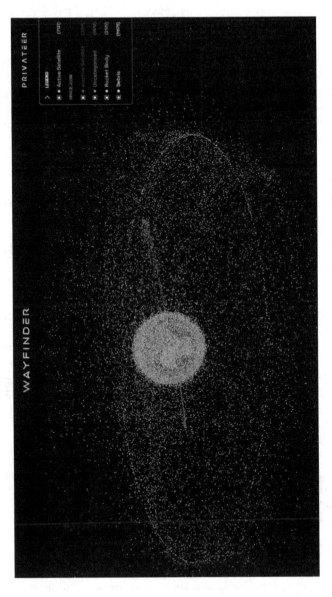

Illustration 6.1 WAYFINDER (*Source* www.privateer.com) (Color figure online)

Saturn's strong gravity, resulting in billions of small chunks of ice and rock coated with dust and other materials (NASA, 2022a), Earth's emerging rings compile human-made objects, consisting of (in)active satellites (see green and dark purple dots in Illustration 6.1), rocket body materials (salmon pink dots), other debris (grey dots) and uncategorised objects (light purple dots). In other words, the human species has managed to create its own orbital ring system—a human-made celestial phenomenon that consists mostly of junk.

Our planet's space junk discs, also referred to as rings of debris in near-Earth space (Barmin et al., 2014), are a cosmic example of how our species' global throwaway society has gone stellar (cf. Hellmann & Luedicke, 2018). The main concern here on Earth about space debris, though, is how such space junk is falling from the sky and the lethal impact it could have on Earth (Condon, 2022). While the likelihood of casualties may seem small, the potential risk of casualties resulting from uncontrolled returning space junk, such as rocket re-entries and falling rocket parts, will increase over the next decade. The southern hemisphere especially is at higher risk than the northern (Byers et al., 2022). The same concerns exist over increased risk of impact by space junk in human-friendly environments in space (Greenbaum, 2020), such as the International Space Station (ISS), which is also itself creating space debris (Johnson & Klinkrad, 2009).

It is not just in space, but also on other celestial bodies, especially Mars and the Moon, that we leave waste behind. Due to 14 different Mars missions, the red planet is littered with 15.694 pounds of human trash from 50 years of robotic Mars explorations (Kilic, 2022). Spacecraft arriving on Mars eject equipment onto the Martian surface, including discarded hardware, inactive spacecraft and crashed spacecraft (id.). Numbers for the Moon are even higher, and given the planned lunar human settlement missions, it is expected that space junk on the Moon will add to 60 + years of lunar debris (Peng, 2015; Witze, 2022).

There are calls for global collective action to make sure that all rocket launching states adopt solutions, such as thoughtful mission design and advanced technologies to reduce the risk of uncontrolled re-entry of spacecraft debris (Byers et al., 2022). These solutions also involve space debris reduction and removal methods (cf. Kaplan, 2009; Mark & Kamath, 2019) ranging from spaced-based laser systems (Shen et al., 2014) to producing reusable spacecraft, such as the Space Rider initiative (Parsonson, 2023).

Clearly, the main concerns of space crime and harms are those about space junk threatening the human species, whereas human-made space junk as a problem in and of itself is receiving less scrutiny (cf. Lampkin, 2021). As a matter of fact, the idea that human beings themselves are a harm in and for space fundamentally opposes the earlier described positive "space is for all humankind" ethos that dominates the spacefaring community and industry (Billings, 2007; Dickens & Ormrod, 2008; Popper & Rakotoniaina, 2019; Vdovychenko, 2022).

In fact, we not only pollute space and other celestial bodies; we also exploit space and infect it with our mere presence. Therefore, secondly, space harm is inflicted by the human species simply by being in space and on other planets due to emissions from space travel (and tourism) and the use of space for anthropocentric purposes, also space mining for energy and minerals (Lampkin, 2021).

Before getting into these exploitative types of space crime and harms, it must be understood that space is not a lawless domain. There are various legal-political agreements, as well as questions and dilemmas regarding who on Earth owns or should own, for example, the Moon, and who is allowed to settle there to "colonize" it (cf. Pop, 2009; Sundahl et al., 2021) by setting up camp, such as the Artemis Base Camp from where Moon-to-Mars missions can take place (Edwards & Cichan, 2021). Many accords, agreements, declarations, governance, international (property) laws, principles and treaties exist, such as the Outer Space Treaty (OST) and ISS Intergovernmental Agreement (IGA). They form a body of space law that regulates how to engage in space (United Nations Office for Outer Space Affairs, 2023). They also tell how the Moon and other celestial bodies ought to be treated. Still, despite international regulations, national regulations and intentions of space-capable nations still seem to adopt a colonialist "first-come, first-served" principles and Rule of Capture-attitude (cf. Dekema, 2022).

Similar to how there were once laws that defined rules of engagement during European colonisation, leading to justification of the plundering of other continents (Mattei & Nader, 2008), it is again law (but more recently) that allows for a silent and slithering colonisation outside Earth (Blomfield, 2004) through

> ...geopolitics, or politics of verticality, and the space arms race [that] are not just about protecting countries, it is protecting asteroids' resources and minerals for mining, and soon, the planets laid stake to by countries

and/or corporations, regardless of the Treaty on Principles Governing the Activities of States in the Exploration and Use of Outer Space, including the Moon and Other Celestial Bodies. (Rothe & Collins, 2023: 10)

Therefore, instead of focusing on what we ought to do legally, it is what humankind is actually doing on (and to) the Moon and Mars, and space in general, in terms of (unconscious) exploitation and annihilation, that is of specific interest here.

Space is being exploited for many reasons. One of the major exploitative undertakings of the human species that is causing tremendous damage to extraplanetary surfaces is that of (commercial) space mining (Johnson, 2020). It can be considered space harm (Loder, 2018), manifested in lunar and Martian eco-theft (Durrani, 2018; Milligan, 2015). For example, and backed by Jeff Bezos and Elon Musk, Rolls Royce is developing a nuclear reactor which is deemed necessary for space mining the Moon and Mars due to the limitations of solar power at such distances (Papadopoulos, 2021). These are examples of very harmful "extreme energy extraction", making, for example, the Moon interesting to private companies for the significant amounts of Helium-3 that can be retrieved from it. Helium-3 is used for energy generation from nuclear fusion, which is a type of extraction that has severe damaging effects on lunar environments (cf. Lampkin, 2021: 246–247). It has been argued that asteroid mining and mining Mars' two moons will also become possible (Hein et al., 2018). For instance, the asteroid-mining start-up AstroForge Inc. plans to launch (at the time of writing) its first two missions to space in April 2023 to extract and refine metals from deep space by refining platinum from a sample of asteroid-like material (Grush, 2023).

While the general public tends to have negative feelings toward mining on Earth's ocean floor, they generally have positive attitudes toward mining on the Moon and asteroids. However, they are concerned about the potential harm that mining could cause to the pristine extraplanetary environment. One reason for the overall positivity toward lunar and asteroid mining is that it is still a relatively abstract concept, and the true impact is not fully conceivable for researchers and participants (Hornsey et al., 2022).

Mining on the Moon and Mars is very harmful (Lampkin, 2021), and perhaps even more harmful than imagined so far. What is problematic about specifically mining asteroids is that it has a far more galactic and existential effect on (potential) life throughout the solar system and Milky

Way than seems to be currently realized. Carnahan et al. (2022) studied how asteroid and comet collisions could potentially seed the oceans of Jupiter's moon Europa with the building blocks necessary for life. An asteroid or comet impact that is capable of reaching halfway through Europa's ice shell would generate heated meltwater that sinks through the rest of the ice. It would alter crater morphology and affect cryovolcanism, which could possibly contribute to the habitability of oceans within icy worlds. This implies that an asteroid and cometary impact could eventually bring about life on Europa, which may also be the case for Saturn's ice moon Enceladus, which also has a liquid ocean beneath its icy crust (Martin et al., 2023).

The idea that asteroids and comets seeded Earth with life is a widely discussed theory in the scientific community; it is still a topic of debate and further research is required to confirm this. Still, research indicates that asteroids and comets seem to have been required once before when they collided with Earth and seeded our own planet with life (Kaiser et al., 2013). Carl Sagan was serious when he said: 'We are all made of star-stuff' (1980). So, by having Earth defence systems that keep our planet safe from asteroids and comets that can crash into our world, we have started to control the course of large and moving celestial bodies. For example, on 26 September 2022, the Double Asteroid Redirection Test (DART) Mission had to impact the Dimorphos, a minor moon of the asteroid Didymos to test the possibility of deflecting an asteroid's trajectory in case of a potential impact (NASA, 2022c).

In line with Garland (2012), it could be argued that such asteroid deflection tests and having an Earth defence system in place uncover a culture of galactic control. However, it is not just about protecting Earth from asteroids and comets; we are also hunting them. NASA's Near-Earth Object (NEO) Surveyor asteroid hunter is designed to change asteroids' and comets' motion in space and their trajectory, far away from Earth (NASA, 2022b; Patton, 2022). This implies that the human species wants to wield the power of influencing celestial bodies, enabling us one day to allow control or destroy life-seeding asteroids and comets. Consequently, we could control life-seeding comets impacting or steering away from, for example, Europa's waters.

Following up on Chapter 4, not only are we a geological force of nature on Earth (UNESCO, 2018) and the metaphorical asteroid that has caused and accelerated the sixth mass extinction on Earth (cf. Francis, 2017; Kolbert, 2014); the human species has the technological ambition (and eventually capacity) to control asteroids and comets that can initiate and also prevent cycles of new life (and mass extinctions) from happening at all on other planets. Controversially, whereas we are a metaphorical asteroid that impacted our own planet, we control real asteroids for life on other planets. All the while we are exploring other planets to live and look on for sentient life, or, "search for extraterrestrial intelligence" (SETI).

6.4   Finding Extraterrestrial Life and Living Extraterrestrially

> You know what I love most about Mars? They still dream. We gave up. They're an entire culture dedicated to a common goal, working together as one to turn a lifeless rock into a garden. We had a garden and we paved it. (Franklin Degraaf in *The Expanse*, Season 1, Episode 3, "Remember the Cant", 2015)

The quote from the sci-fi series *The Expanse* shows how in the future on Earth we destroyed our habitable environment, having it changed into a desert, a rock-formed planet, while on Mars we are trying to terraform the red rocky planet into a second green and blue Earth. The character's wish triggers the question: what if the human species had tried to keep Earth safe and better habitable, investing its interest and technology in and on Earth, instead of focusing on trying to find life and living possibilities off Earth? This is not merely a sci-fi question, but is becoming a realistic concern as the human species has set its sights on the stars to find other life and new living places. Along with this will come human-caused harm to other planets and potentially other life forms out there (Alberro, 2022; Bohlander, 2021; Deudney, 2020; Lampkin, 2021; Miller, 2001; Mitchell et al., 2018; Takemura, 2019).

NASA is currently developing an improved version of the James Webb Telescope, provisionally titled the Habitable Worlds Observatory. Its mission is to search for life on planets like ours in about seven years (Clery, 2023 – online source). Additionally, around 2027, NASA's 'dark energy and exoplanet-hunting Nancy Grace Roman Observatory' (id.) will commence its mission. This search for extraterrestrial intelligence (SETI) and Earth-like planets is often popularly referred to as "alien hunt" or "hunting exoplanets", also by space agencies such as NASA (Chu, 2017). There is even a "hunt" for entire other star systems like our solar system, of which the Alpha Centauri star system is the most keenly considered. In particular sending laser-powered micro ships is a novel but remarkable new type of technology to go to these star systems (Loughran, 2016). One such initiative, the Breakthrough Starshot project, entails sending an ultra-lightweight spacecraft across that is able to cover vast distances of four lightyears with unparalleled speed, reaching its destination within 20 years (Peretz et al., 2022). Moreover, with the help of Earth-based observatories, there is the ambition to discover more easily advanced alien life forms and their possible activity of hypothetical warp drives, as has been suggested by a team of scientists (Sellers et al., 2022). By the way, warp drives are one of the key features of the *Star Trek* spaceships (Krauss, 2007), showing how sci-fi is becoming real. These instances serve as examples of the human species' desire to explore and find habitable environments beyond Earth.

There are various signs that life has existed on other planets, not only on Mars, but also elsewhere, including signs that indicate that life elsewhere may start to exist in the future when we aim to send human missions to Mars, for example (cf. Hellweg & Baumstark-Khan, 2007; Magnuson et al., 2022). Recently, approximately 71 trillion gallons of water in the form of microscopic beads of silicate were discovered on the lunar surface (He et al., 2023). Water is essential for human and other life as we know it. Much water on the Moon is also key to space inhabitation and space exploration (cf. Levchenko et al., 2021). There are various suggestions that relict glaciers and dried-up lake beds on Mars may contain opal gemstones, as NASA's Mars Curiosity rover indicated. These opals are evidence that water and rock were interacting underneath Mars' surface more recently than was previously thought, which suggests microbial life may have lived there (Gabriel et al., 2022; Lee et al., 2023). It has also been suggested that Martian tsunamis brought about microbial life (Veysi, 2022). The presence of salty lakes and ice on Mars is another indication of potential previous life (Joseph et al., 2022; Tung et al., 2005).

The quest for life goes beyond the Moon and Mars. On Venus (cf. Greaves et al., 2021), on Jupiter's moons (Salter et al., 2022), on asteroids (Oba et al., 2023) and in other (much older) star systems like our solar system (Elms et al., 2022), even in the vacuum of space itself (Tobin et al., 2023), there have been (strong) indicators of (previous) microbial life forms and of Earth-like planets that existed long before our world. Recent research has indicated that low metallicity star systems are more likely to host extraterrestrial complex life forms (Shapiro et al., 2023). It suggests that planets orbiting stars with a low concentration of heavy elements could be more habitable due to a protective ozone layer shielding them from UV radiation.

Not just by "hunting" extraterrestrial life using telescopes and atmospheric modelling but also by retrieving samples from, for example, the Moon and Mars (NASA, 2023a), we hope to discover extraterrestrial life. However, there are worries about the possibility that rock samples brought back from Mars may contain Martian germs that could escape into our planet's environment, potentially causing problems or diseases (cf. Bianciardi, 2022; Stuart, 2022). The possibility that Earth's micro-bacteria could contaminate Mars organically through the presence of humans and/or human-made objects (cf. Greenfieldboyce, 2022), such as the Mars Rover, is not that big a worry, despite the fact that it should be (Lampkin & Wyatt, 2022; Van Houdt et al., 2012).

Finally, one of the main reasons why the human species is searching for extraterrestrial life and Earth-like environments outside our own planet is to one day live off Earth. Moon bases, such as the Artemis Base Camp, are expected to be ready for human occupation by 2025 (Creech et al., 2022). Elon Musk has spoken about his vision for human beings to become a multiplanetary species, with settlements on both Earth and Mars (2017). Another reason why humans living on other planets is becoming a reality is the technological development of making travel to Mars faster and shorter, up to 45 to 39 days now (NASA, 2023b; Chen, 2021). Technology is also advancing sustainable living in space. In preparation for "the colonisation" of other planets, Interstellar Lab, a French-American start-up that develops sustainable living systems on Earth, is designing environmentally controlled, closed-loop systems for life on other planets. Their goal is to produce and recycle almost everything, making water, air and food as renewable as possible for human life in total autonomy (3ds.com, 2023). There is even talk of terraforming other planets to the likeness of Earth and how we imagine it to be (Pak, 2016; Persinger, 2020). In addition to tech billionaires and well-funded international start-ups, crowd-sourcing online communities such as Asgardia Space Nation are actively recruiting members to join and contribute to living in space, data storage in space and bio-engineered space babies, to create, they claim, a better future through democracy and innovation (Asgardia.space, 2023).

The idea that more human beings are going to float around for a short period of time in space due to space commercial flights (Loizou, 2006), or for a longer time in (bigger) space stations and commercial space hotels (cf. Cottier, 2021), implies an increase of substantial human and harmful activity in space. The human species' expansion into space is driven by 'a discourse of multiplanetarity that veils the continuity of extractivist capitalism/colonialism in a narrative of futurity and progress' (Temmen, 2022: 477). In other words, (narratives on) space colonisation and terraforming of whichever extraterrestrial body function as a distraction from Earth's climate change, pretending humanity will overcome the sixth mass extinction elsewhere. However,

> …[t]hese logics, which draw from myths of American colonialism and the "western frontier" […] imagining that, through technological intervention and expansion, the root causes of climate change do not actually need to be addressed because we can simply offset damage to the Earth through the acquisition and transformation of new planets/moons/asteroids. (Persinger, 2020: 1)

That neocolonial, expansionist and exploitative crusade into space is driving and making the human species use (more) AI-powered tools and technologies, as well as human enhancement (see Chapter 5). In fact, from a moral point of view, it is argued that for

> …any space program involving long-term human missions […] there is a need for new, more radical solutions [such as] human enhancement and a fully automated space exploration, based on highly advanced AI [for which] there are strong reasons to consider human enhancement, including gene editing of germ line and somatic cells, as a moral duty. (Szocik, 2020: 122)

There are AI-designed spacecraft and rovers for exploration of planets to assist space exploration missions and building Moon bases, such as the Artemis Mission (ESA, 2023; Wodecki, 2023). AI is also used to make more (accurate) discoveries in space. By applying an algorithm learning programme, called XGBoost, a large number of so-called Milky Way fossil stars have been discovered, giving an insight into the Milky Way's "poor old heart" of formation and disk galaxies that would have remained invisible otherwise (Rix et al., 2022).

We also train and human-enhance ourselves and our performance for living on Mars in Mars-like environments on Earth (Terhorst & Dowling, 2022), a place where our biology is not meant to be. On Mars, for example, we can live for a maximum of four years, after which we risk being affected by particle radiation from the Sun and other distant stars and even galaxies (Dobynde et al., 2021). Experts who are looking into the future have come up with the idea of changing the physiology of astronauts to better protect them from radiation and the effects of weightlessness, which is crucial for human Mars missions (Gambacurta et al., 2019; Li et al., 2017; McKie, 2023). There have been attempts to make astronauts resilient to harmful microgravitational effects on the body, including the deterioration of bone tissue (osteoporosis), by taking medication that reduces osteoporosis (cf. Leblanc et al., 2013; Smith, 2020). However, human enhancement medicines like these could alter the human body to such an extent that it becomes harder, if not impossible, to return to Earth because the altered bone structure would no longer be able to resist Earth's gravity (Szocik et al., 2019). Space-based human enhancement also entails DNA splicing. For example, Chinese military research suggests that it could be feasible to insert the genes of tardigrades, which are among the toughest animals known to humankind, into human DNA to enhance a person's resistance to radiation from the Sun and perhaps other ailments (Hashimoto et al., 2016).

Like the human enhancement and AI technologies aiming for singularity on Earth that were discussed in Chapter 5, such upgrading and AI-fication of the human species goes even further in and because of our space expansionism (cf. Beames, 2022). Whether it is possible or not, these endeavours to enhance reflect the ambitions of the human species that is willing 'to change human nature if we're going to survive' in space, as environmental scientist Carmel Johnston stated, having lived in an Earth-based replica environment of Mars (Knight, 2017 – online source). With the ambition to explore off Earth, the human species imagines a life in space is possible if it becomes more dependent on AI and alters its biology, morally justifying yet further redesign of itself. So, we are a self-redesigning species and a fast self-evolving species that is not only going away from its natural biological habitat into space, but is also evolving away from it in space, because of its aspirations of extraterrestrial livelihoods. We are willing to achieve singularity more quickly and change our human biology more rigorously for space exploration and inhabitation, which leads to the paradox that we are willing to alter our humanness drastically into something not-so-human "for all humankind" in space.

6.5 Conclusion

From an economic justice perspective, the human species is made of Earth elements – elements coming from anywhere in the universe, which makes us 'galactic beings as well and [we] are mysteriously related to the entire universe' (Hartzok, 2001: 7). Although the human species has not created Earth, it may very well be its destroyer due to the self-inflicted Anthropocene mass extinction, which is being denied (see Chapter 4). Therefore, imaginably, the human species as a "galactic being" that aims to inhabit space and other planets in the future, is prepared to alter itself even more into something not so human, while causing (and denying) cosmic harm through mining activities (cf. Munévar, 2014; Noriyoshi, 2019; Walcott, 2014), and (unknowingly) contaminating, even killing, extraterrestrial life on those planets (Alberro, 2022; Bohlander, 2021). So, 'if the analog of human global expansion is a reliable guide, solar expansion will be chronically violent and thus less desirable than advocates believe' (Deudney, 2020: 346). Then, space is where the destructive nature of the human species becomes apparent as well (Huebert & Block, 2006; Lampkin, 2021); not just our (mass) exploitative tendencies, but also our total annihilative tendencies could be extended to extraterrestrial genocide and (human-caused and -accelerated) interplanetary mass extinctions.

References

3ds.com. (2023). Sustainable life on Earth and beyond. https://www.3ds.com/insights/customer-stories/interstellar-lab-sustainable-life-on-earth

AGU/American Geophysical Union. (2017–2019). Earth and Space Science is Essential for Society. https://agupubs.onlinelibrary.wiley.com.doi/toc/; https://doi.org/10.1002/(ISSN)2333-5084.SCISOC1

Alberro, H. (2022). HG Wells, earthly and post-terrestrial futures. *Futures, 140,* 102954.

Asgardia.space. (2023). The space nation. https://asgardia.space/

Axpe, E., Chan, D., Abegaz, M. F., Schreurs, A. S., Alwood, J. S., Globus, R. K., & Appel, E. A. (2020). A human mission to Mars: Predicting the bone mineral density loss of astronauts. *PLoS ONE, 15*(1), e0226434.

Baker, M. (2020). NASA Astronaut Anne McClain accused by spouse of crime in space. https://www.nytimes.com/2019/08/23/us/astronaut-space-invest igation.html

Barmin, I. V., Dunham, D. W., Kulagin, V. P., Savinykh, V. P., & Tsvetkov, V. Y. (2014). Rings of debris in near-Earth space. *Solar System Research, 48*(7), 593–600.

Beames, C. (2022). AI in space and its future use in warfare. https://www.for bes.com/sites/charlesbeames/2022/12/21/ai-in-space-and-its-future-use-in-warfare/amp/

Bianciardi, G. (2022). Is life on Mars a danger to life on Earth? NASA's Mars sample return. *Journal of Astrobiology, 11*, 14–20.

Billings, L. (2007). Ideology, advocacy, and spaceflight: Evolution of a cultural narrative. In S. J. Dick & R. D. Launius (Eds.), *Societal impact of spaceflight* (pp. 483–499). NASA.

Billings, L. (2017). Should humans colonize other planets? *Theology and Science, 15*(3), 321–332.

Blomfield, H. D. (2004). Frontier society: Perpetuation and misrepresentation of humankind in outer space policy. https://open.library.ubc.ca/soa/cIRcle/collections/ubctheses/831/items/1.0091574

Bohlander, M. (2021). Metalaw-What is it good for? *Acta Astronautica, 188*, 400–404.

Bradley, A. M., & Wein, L. M. (2009). Space debris: Assessing risk and responsibility. *Advances in Space Research, 43*(9), 1372–1390.

Brown, T. (2022). Can Starlink satellites be lawfully targeted? https://lieber.wes tpoint.edu/can-starlink-satellites-be-lawfully-targeted/

Byers, M., Wright, E., Boley, A., & Byers, C. (2022). Unnecessary risks created by uncontrolled rocket reentries. *Nature Astronomy, 6*(9), 1093–1097.

Cannon, K. M., & Britt, D. T. (2019). Feeding one million people on Mars. *New Space, 7*(4), 245–254.

Carnahan, E., Vance, S. D., Cox, R., & Hesse, M. A. (2022). Surface-to-ocean exchange by the sinking of impact generated melt chambers on Europa. *Geophysical Research Letters* (p. e2022GL100287).

Chatzipanagiotis, M. (2011). *The legal status of space tourists in the framework of commercial suborbital flights.* Carl Heymanns Verlag.

Chen, S. (2021). How China's space station could help power astronauts to Mars. https://www.scmp.com/news/china/science/article/3135770/how-chinas-space-station-could-help-power-astronauts-mars

Christensen, C., Starzyk, J., Potter, S., & Boensch, N. (2017). Start-up space: Update on investment in commercial space ventures. https://brycetech.com/downloads/Bryce_Start_Up_Space_2017.pdf

Chu, J. (2017). Scientists make huge dataset of nearby stars available to public. https://exoplanets.nasa.gov/news/1413/scientists-make-huge-dat aset-of-nearby-stars-available-to-public/

Clark, J., Koopmans, C., Hof, B., Knee, P., Lieshout, R., Simmonds, P., & Wokke, F. (2014). Assessing the full effects of public investment in space. *Space Policy, 30*(3), 121–134.

Clery, D. (2023). NASA unveils initial plan for multibillion-dollar telescope to find life on alien worlds. https://www.science.org/content/article/nasa-unv eils-initial-plan-multibillion-dollar-telescope-find-life-alien-worlds

Cohen, S. (2013). *States of denial: Knowing about atrocities and suffering*. John Wiley & Sons.

Condon, S. (2022). Space junk is falling from the sky. We are still not doing enough to stop it. https://www-zdnet-com.cdn.ampproject.org/c/s/www. zdnet.com/google-amp/article/space-junk-is-falling-from-the-sky-we-are-still-not-doing-enough-to-stop-it/

Cottier, C. (2021). The first 'Space hotel' plans to open in 2027. https://astron omy.com/news/2021/11/the-first-space-hotel-plans-to-open-in-2027

Creech, S., Guidi, J., & Elburn, D. (2022). Artemis: An overview of NASA's activities to return humans to the Moon. *2022 IEEE Aerospace Conference (AERO)* (pp. 1–7).

Dass, R. S. (2017). The high probability of life on Mars: A brief review of the evidence. *Cosmology, 27*, 62–73.

Dekema, B. M. (2022). To infinity and beyond: Shifting the space regula-tory framework to create conservation-minded expansion. *Natural Resources Journal, 62*(2), 237–256.

Deudney, D. (2020). *Dark skies: Space expansionism, planetary geopolitics, and the ends of humanity*. Oxford University Press.

Dickens, P., & Ormrod, J. (2008). Who really won the space race? *Monthly Review, 59*(9), 30–37.

Diederiks-Verschoor, I. (1967). Some observations regarding the treaty on space law. *Proceedings on the Law of Outer Space, 10*, 164–169.

Diederiks-Verschoor, I. (1979). Space law as it affects domestic law. *Journal of Space Law, 7*(1), 39–46.

Dobynde, M. I., Shprits, Y. Y., Drozdov, A. Y., Hoffman, J., & Li, J. (2021). Beating 1 sievert: Optimal radiation shielding of astronauts on a mission to Mars. *Space Weather, 19*(9), e2021SW002749.

Dunnett, O., Maclaren, A., Klinger, J., Maria, K., Lane, D., & Sage, D. (2019). Geographies of outer space: Progress and new opportunities. *Progress in Human Geography, 43*(2), 314–336.

Durrani, H. A. (2018). Interpreting space resources obtained: Historical and postcolonial interventions in the law of commercial space mining. *Columbia Journal of Transnational Law, 57*, 403–460.

Edwards, C. M., & Cichan, T. (2021). From the Moon to Mars base camp: An updated architecture that builds on artemis. *ASCEND 2021* (p. 4137).

Elms, A. K., Tremblay, P. E., Gänsicke, B. T., Koester, D., Hollands, M. A., Gentile Fusillo, N. P., Cunningham, T., & Apps, K. (2022). Spectral analysis of ultra-cool white dwarfs polluted by planetary debris. *Monthly Notices of the Royal Astronomical Society, 517*(3), 4557–4574.

Emery, A. J. (2013). *The case for spacecrime: The rise of crime and piracy in the space domain.* https://apps.dtic.Mil/sti/pdfs/ADA615000.pdf

Erwin, S., & Werner, D. (2022). Dark clouds, silver linings: Five ways war in Ukraine is transforming the space domain. https://spacenews.com/dark-clouds-silver-linings-five-ways-war-in-ukraine-is-transforming-the-space-domain/

ESA. (2022). Space and daily life. https://www.esa.int/Science_Exploration/Human_and_Robotic_Exploration/Education/Space_and_daily_life

ESA. (2023). Guide morphing rovers across alien world in evolutionary computing contest. https://www.esa.int/Enabling_Support/Space_Engineering_Technology/Guide_morphing_rovers_across_alien_world_in_evolutionary_computing_contest

Eski, Y. (2011). 'Port of call': Toward a criminology of port security. *Criminology & Criminal Justice, 11*(5), 415–431.

Eski, Y. (2023). Crime among the stars: Space criminology and the domain of limitlessness (in Dutch: Criminaliteit Tussen de Sterren: Ruimtecriminologie en het Domein van het Grenzeloze). *De Criminoloog, 28,* 7. https://research.vu.nl/files/233723169/Nieuwsbrief28_def.pdf

European Commission. (2023). An EU Space Strategy for Security and Defence to ensure a stronger and more resilient EU. https://ec.europa.eu/commission/presscorner/detail/en/ip_23_1601

Foust, J. (2023). White House proposes $27.2 billion for NASA in 2024. https://spacenews.com/white-house-proposes-27-2-billion-for-nasa-in-2024/

Francis, M. R. (2017). The sixth mass extinction: We aren't the dinosaurs, we're the asteroid. https://www.thedailybeast.com/the-sixth-mass-extinction-we-arent-the-dinosaurs-were-the-asteroid

Froehlich, A. (2021). *Assessing a Mars agreement including human settlements.* Springer.

Gabriel, T. S., Hardgrove, C., Achilles, C. N., Rampe, E. B., Rapin, W. N., Nowicki, S., Czarnecki, S., Thompson, L., Nikiforov, S., Litvak, M., Mitrofanov, I., & McAdam, A. (2022). On an extensive late hydrologic event in Gale Crater as indicated by water-rich fracture halos. *Journal of Geophysical Research: Planets, 127*(12), e2020JE006600.

Gambacurta, A., Merlini, G., Ruggiero, C., Diedenhofen, G., Battista, N., Bari, M., Balsamo, M., Piccirillo, S., Valentini, G., Mascetti, G., & Maccarrone, M. (2019). Human osteogenic differentiation in Space: Proteomic and epigenetic clues to better understand osteoporosis. *Scientific Reports, 9*(1), 8343.

Garland, D. (2012). *The culture of control: Crime and social order in contemporary society*. University of Chicago Press.

Gohd, C. (2020). Astronaut Anne McClain's estranged wife charged with lying about alleged 'space crime.' https://space.com/astronaut-anne-mcclain-wife-charged-lying-space-crime.html

Gorove, S. (1972a). Criminal jurisdiction in outer space. *The International Lawyer, 6*(2), 313–323.

Gorove, S. (1972b). Pollution and outer space: A legal analysis and appraisal. *New York University Journal of International Law and Politics, 5*(1), 53–66.

Grabosky, P. N., & Smith, R. G. (1998). *Crime in the digital age: Controlling telecommunications and cyberspace illegalities*. Transaction Publishers.

Greaves, J. S., Richards, A., Bains, W., Rimmer, P. B., Sagawa, H., Clements, D. L., Seager, S., Petkowski, J. J., Sousa-Silva, C., Ranjan, S., Drabek-Maunder, E., & Hoge, J. (2021). Phosphine gas in the cloud decks of Venus. *Nature Astronomy, 5*(7), 655–664.

Greenbaum, D. (2020). Space debris puts exploration at risk. *Science, 370*(6519), 922–922.

Greenfieldboyce, N. (2022). NASA is bringing rocks back from Mars, but what if those samples contain alien life? https://www.npr.org/2022/05/04/109 5645081/nasa-is-bringing-rocks-back-from-mars-but-what-if-those-samples-contain-alien-li

Gregg, J. (2021). Space Biz. *The cosmos economy* (pp. 135–144). Copernicus.

Grush, L. (2023). Asteroid-mining startup AstroForge to launch first space missions this year. https://www.bloomberg.com/news/articles/2023-01-24/asteroid-mining-startup-astroforge-plans-first-platinum-refining-space-mis sions

Gunasekara, S. G. (2010). The march of science: Fourth amendment implications on remote sensing in criminal law. *Journal of Space Law, 36*(1), 115–142.

Hartzok, A. (2001). Democracy, earth rights, and next economy. Paper presented at the 21st annual E. F. Schumacher lectures. http://www.earthrights.net/docs/schumacher.html

Hashimoto, T., Horikawa, D. D., Saito, Y., Kuwahara, H., Kozuka-Hata, H., Shin-i, T., Minakuchi, Y., Ohishi, K., Motoyama, Y., Aizu, T., & Kunieda, T. (2016). Extremotolerant tardigrade genome and improved radiotolerance of human cultured cells by tardigrade-unique protein. *Nature Communications, 7*(1), 12808.

Haughney, E. W. (1963, June 18–20). Criminal responsibility in outer space. Paper presented at the Conference on Space Science and Space Law, University of Oklahoma, Norman, Oklahoma.

He, H., Ji, J., Zhang, Y., Hu, S., Lin, Y., Hui, H., Hao, J., Li, R., Yang, W., Tian, H., Zhang, C., & Wu, F. (2023). A solar wind-derived water reservoir on the Moon hosted by impact glass beads. *Nature Geoscience* (pp. 1–7).

Hein, A. M., Saidani, M., & Tollu, H. (2018). Exploring potential environmental benefits of asteroid mining. ArXiv:1810.04749.

Hellmann, K. U., & Luedicke, M. K. (2018). The throwaway society: A look in the back mirror. *Journal of Consumer Policy, 41*, 83–87.

Hellweg, C. E., & Baumstark-Khan, C. (2007). Getting ready for the manned mission to Mars: The astronauts' risk from space radiation. *Naturwissenschaften, 94*(7), 517–526.

Hoffman, J. A., Hecht, M. H., Rapp, D., Hartvigsen, J. J., SooHoo, J. G., Aboobaker, A. M., McClean, J. B., Liu, A. M., Hinterman, E. D., Nasr, M., & Hariharan, S. (2022). Mars Oxygen ISRU Experiment (MOXIE)—Preparing for human Mars exploration. *Science Advances, 8*(35), 1–6.

Hornsey, M. J., Fielding, K. S., Harris, E. A., Bain, P. G., Grice, T., & Chapman, C. M. (2022). Protecting the planet or destroying the universe? *Understanding reactions to space mining. Sustainability, 14*(7), 4119.

Huebert, J. H., & Block, W. (2006). Space Environmentalism, Property Rights, and the Law. *University of Memphis Law Review, 37*, 281.

IPCC. (2021). Climate change 2021: The physical science basis report. https://www.ipcc.ch/report/ar6/wg1/downloads/report/IPCC_AR6_WGI_SPM.pdf

Jackson, T. (2022). Billionaire space race: The ultimate symbol of capitalism's obsession with growth. https://nextnature.net/magazine/story/2022/billionaire-space-race-the-ultimate-symbol-of-capitalism-s-obsession-with-growth

Johnson, M. R. (2020). Mining the high frontier: Sovereignty, property and humankind's common heritage in outer space (Doctoral dissertation). http://hdl.handle.net/10453/142380

Johnson, N., & Klinkrad, H. (2009). The International space station and the space debris environment: 10 Years on. *5th European Conference on Space Debris* (No. JSC-CN-17944). https://ntrs.nasa.gov/citations/20090004997

Joseph, R. G., Gibson, C., Wolowski, K., Bianciardi, G., Kidron, G. J., Armstrong, R. A., & Schild, R. (2022). Evolution of life in the oceans of Mars? Episodes of global warming, flooding, rivers, lakes, and chaotic orbital obliquity. *Journal of Astrobiology, 13*(2022), 14–126.

Kaiser, R. I., Stockton, A. M., Kim, Y. S., Jensen, E. C., & Mathies, R. A. (2013). On the formation of dipeptides in interstellar model ices. *The Astrophysical Journal, 765*(2), 111.

Kaku, M. (2023). Michio Kaku: 3 mind-blowing predictions about the future. https://bigthink.com/the-well/predictions-for-the-future/

Kaplan, M. (2009). Survey of space debris reduction methods. In *AIAA space 2009 Conference & Exposition* (p. 6619).

Kern, S. (2021). No, billionaires won't "escape" to space while the world burns. https://salon.com/2021/07/07/no-billionaires-wont-escape-to-space-while-the-world-burns/

Kilic, C. (2022). Mars is littered with 15,694 pounds of human trash from 50 years of robotic exploration. https://theconversation.com/mars-is-littered-with-15-694-pounds-of-human-trash-from-50-years-of-robotic-exploration-188881

Knight, A. (2017). This woman spent a year 'on Mars' as a psychological experiment. https://www.vice.com/en/article/zmb4q3/this-woman-spent-a-year-on-mars-as-a-psychological-experiment

Kolbert, E. (2014). *The sixth extinction: An unnatural history.* A&C Black.

Kotzé, J., & Antonopoulos, G. A. (2022). Con Air: Exploring the trade in counterfeit and unapproved aircraft parts. *The British Journal of Criminology,* 1–16.

Krauss, L. M. (2007). *The physics of Star Trek.* Basic Books.

Kuipers, A. (2022, September). Galactic entrepreneurship. *Holland Herald* (p. 22).

Lampkin, J. (2020). Mapping the terrain of an astro-green criminology: A case for extending the green criminological lens outside of planet Earth. *Astropolitics, 18*(3), 238–259.

Lampkin, J. (2021). Should criminologists be concerned with outer space?: A proposal for an 'Astro-criminology'. https://research.leedstrinity.ac.uk/en/publications/should-criminologists-be-concerned-with-outer-space-a-propos al-fo

Lampkin, J., & White, R. (2023). *Space criminology: Analysing human relationships with outer space.* Springer/Palgrave Macmillan.

Lampkin, J., & Wyatt, T. (2022). Widening the scope of "Earth" jurisprudence and "Green" criminology? Toward preserving extra-terrestrial heritage sites on celestial bodies. In J. Gacek & R. Jochelson (Eds.), *Green criminology and the law* (pp. 309–329). Springer.

Lasch, C. (1979). *The culture of narcissism.* Norton.

Leblanc, A., Matsumoto, T., Jones, J., Shapiro, J., Lang, T., Shackelford, L., & Ohshima, H. (2013). Bisphosphonates as a supplement to exercise to protect bone during long-duration spaceflight. *Osteoporosis International, 24,* 2105–2114.

Lee, P., Shubham, S., & Schutt, J. W. (2023). A relict glacier near Mars' equator: Evidence for recent glaciation and volcanism in Eastern Noctis Labyrinthus. *54th Lunar and Planetary Science Conference 2023* (LPI Contrib. No. 2806). https://www.hou.usra.edu/meetings/lpsc2023/pdf/2998.pdf

Levchenko, I., Xu, S., Mazouffre, S., Keidar, M., & Bazaka, K. (2021). Mars colonization: Beyond getting there. *Terraforming Mars* (pp. 73–98).

Li, J., Bonkowski, M. S., Moniot, S., Zhang, D., Hubbard, B. P., Ling, A. J., Rajman, L. A., Qin, B., Lou, Z., Gorbunova, V., Aravind, L., & Sinclair, D. A. (2017). A conserved NAD+ binding pocket that regulates protein-protein interactions during aging. *Science, 355*(6331), 1312–1317.

Loder, R. E. (2018). Asteroid mining: Ecological jurisprudence beyond Earth. *Virginia Environmental Law Journal, 36*(3), 275–317.

Logsdon, J. M. (2015). Why did the United States retreat from the moon? *Space Policy, 32*, 1–5.

Loizou, J. (2006). Turning space tourism into commercial reality. *Space Policy, 22*(4), 289–290.

Loughran, J. (2016). Micro spaceships powered by lasers to search for alien life [News Briefing]. *Engineering & Technology, 11*(4), 18–18.

Magnuson, E., Altshuler, I., Fernández-Martínez, M. Á., Chen, Y. J., Maggiori, C., Goordial, J., & Whyte, L. G. (2022). Active lithoautotrophic and methane-oxidizing microbial community in an anoxic, sub-zero, and hyper-saline High Arctic spring. *The ISME Journal* (pp. 1–11).

Marikar, S. (2018). The rich are planning to leave this wretched planet. https://nytimes.com/2018/06/09/style/axiom-space-travel.html

Mark, C. P., & Kamath, S. (2019). Review of active space debris removal methods. *Space Policy, 47*, 194–206.

Martin, E. S., Whitten, J. L., Kattenhorn, S. A., Collins, G. C., Southworth, B. S., Wiser, L. S., & Prindle, S. (2023). Measurements of regolith thicknesses on Enceladus: Uncovering the record of plume activity. *Icarus, 392*, 115369.

Mattei, U., & Nader, L. (2008). *Plunder: When the rule of law is illegal*. John Wiley & Sons.

McKie, R. (2023). How far should we go with gene editing in pursuit of the 'perfect' human? https://www.theguardian.com/science/2023/feb/05/how-far-should-we-go-with-gene-editing-in-pursuit-of-the-perfect-human

Michalowski, R. J., & Kramer, R. C. (2006). *State-corporate crime: Wrongdoing at the intersection of business and government*. Rutgers University Press.

Miller, G. D. (2019). Space pirates, geosynchronous guerrillas, and nonterrestrial terrorists: Nonstate threats in space. *Air & Space Power Journal, 33*(3), 33–51.

Miller, R. W. (2001). Astroenvironmentalism: The case for space exploration as an environmental issue. *Electronic Green Journal, 1*(15), 1–7.

Milligan, T. (2015). *Nobody owns the moon: The ethics of space exploitation*. McFarland.

Mitchell, J., Evans, C., & Stansbery, E. (2018). Next steps in planetary protection for human spaceflight. *42nd COSPAR Scientific Assembly, 42*, PPP-1. https://ntrs.nasa.gov/api/citations/20180004777/downloads/20180004777.pdf

Morgan Stanley. (2020). Space: Investing in the final frontier. https://morganstanley.com/ideas/investing-in-space

Munévar, G. (2014). Space exploration and human survival. *Space Policy, 30*(4), 197–201.

Musk, E. (2017). Making humans a multi-planetary species. *New Space*, 5(2), 46–61.

NASA. (2022a). Saturn. https://solarsystem.nasa.gov/planets/saturn/in-depth/

NASA. (2022b). Construction begins on NASA's next-generation asteroid hunter. https://www.jpl.nasa.gov/news/construction-begins-on-nasas-next-generation-asteroid-hunter

NASA. (2022c). NASA confirms DART mission impact changed asteroid's motion in space. https://www.nasa.gov/press-release/nasa-confirms-dart-mis sion-impact-changed-asteroid-s-motion-in-space

NASA. (2023a). NASA to launch new Mars sample receiving project office at Johnson. https://www.nasa.gov/press-release/nasa-to-launch-new-mars-sam ple-receiving-project-office-at-johnson

NASA. (2023b). New class of bimodal NTP/NEP with a wave rotor topping cycle enabling fast transit to Mars. https://www.nasa.gov/directorates/spa cetech/niac/2023b/New_Class_of_Bimodal/

Noriyoshi, T. (2019). Astro-Green criminology: A new perspective against space capitalism. https://core.ac.uk/reader/211127200

NSR. (2022). Developing moon market propelled by 250+ missions and $105 billion in revenue through decade. https://www.nsr.com/nsr-develo ping-moon-market-propelled-by-250-missions-and-105-billion-in-revenue-thr ough-decade/

Oba, Y., Koga, T., Takano, Y., Ogawa, N.O., Ohkouchi, N., Sasaki, K., Sato, H., Glavin, D. P., Dworkin, J. P., Naraoka, J., Tachibana, S., & Hayabusa2-initial-analysis SOM team. (2023). Uracil in the carbonaceous asteroid (162173) Ryugu. *Nature Communications*, 14(1), 1292.

Ormrod, J. S. (2007). Pro-space activism and narcissistic phantasy. *Psychoanalysis, Culture & Society*, 12(3), 260–278.

Pak, C. (2016). *Terraforming: Ecopolitical transformations and environmentalism in science fiction.* Liverpool University Press.

Papadopoulos, L. (2021). Rolls-Royce is developing a nuclear reactor for mining the Moon and Mars. https://interestingengineering.com/innovation/rolls-royce-nuclear-reactor-for-mining-the-moon-and-mars

Parsonson, A. (2023). A look at the reusable Space Rider project. https://eur opeanspaceflight.substack.com/p/a-look-at-the-reusable-space-rider

Patel, Z. S., Brunstetter, T. J., Tarver, W. J., Whitmire, A. M., Zwart, S. R., Smith, S. M., & Huff, J. L. (2020). Red risks for a journey to the red planet: The highest priority human health risks for a mission to Mars. *NPJ Microgravity*, 6(1), 1–13.

Patton, T. (2022). Construction begins on NASA's NEO surveyor asteroid hunter. *The Journal of Space Commerce*. https://exterrajsc.com/construction-begins-on-nasas-neo-surveyor-asteroid-hunter/2022/12/27/

Peng, B. (2015). Dangers of Space Debris. *Berkeley Scientific Journal, 19*(2).

Peoples, C. (2022). Global uncertainties, geoengineering and the technopolitics of planetary crisis management. *Globalizations, 19*(2), 253–267.

Peoples, C., & Stevens, T. (2020). At the outer limits of the international: Orbital infrastructures and the technopolitics of planetary (in) security. *European Journal of International Security, 5*(3), 294–314.

Peretz, E., Mather, J. C., Hamilton, C., Pabarcius, L., Hall, K., Fugate, R. Q., & Klupar, P. (2022). Orbiting laser configuration and sky coverage: Coherent reference for Breakthrough Starshot ground-based laser array. *Journal of Astronomical Telescopes, Instruments, and Systems, 8*(1), 017004–017004.

Persinger, K. (2020). Constructing reality: An investigation of climate change and the terraforming imaginary. *The Macksey Journal, 1*(1), 1–16.

Pop, V. (2009). *Who owns the moon?* Springer.

Popper, J., & Rakotoniaina, S. (2019, October 21–25). Re-imagining outer space. Paper presented at the 70th International Astronautical Congress (IAC), Washington DC, United States.

Potter, M. (1995). The outer space cyberspace nexus: Satellite crimes. *Journal of Space Law, 23*(1), 55–56.

Rhea Group. (2022). RHEA and SpeQtral to develop quantum-safe link between Singapore and Europe. https://www.rheagroup.com/rhea-and-speqtral-to-develop-quantum-safe-link-between-singapore-and-europe/

Rix, H. W., Chandra, V., Andrae, R., Price-Whelan, A. M., Weinberg, D. H., Conroy, C., Fouesneau, M., Hogg, D. W., De Angeli, F., Naidu, R. P., Xiang, M., & Ruz-Mieres, D. (2022). The poor old heart of the Milky Way. *The Astrophysical Journal, 941*(1), 45.

Robinson, G. S. (1974). Psychoanalytic techniques supporting biojuridics in space. *Journal of Space Law, 2*(1), 95–106.

Rothe, D. L., & Collins, V. E. (2023). Planetary geopolitics, space weaponization and environmental harms. *The British Journal of Criminology, azad003*, 1–16.

Sagan, C. (1980). *Cosmos* (mini-series). PBS television.

Salter, T. L., Magee, B. A., Waite, J. H., & Sephton, M. A. (2022). Mass spectrometric fingerprints of bacteria and archaea for life detection on icy moons. *Astrobiology, 22*(2), 143–157.

Sellers, L., Bobrick, A., Martire, G., Andrews, M., & Paulini, M. (2022). Searching for intelligent life in gravitational wave signals part I: Present capabilities and future horizons. ArXiv: 2212.02065.

Shammas, V. L., & Holen, T. B. (2019). One giant leap for capitalistkind: Private enterprise in outer space. *Palgrave Communications, 5*(1), 1–9.

Shapiro, A. V., Brühl, C., Klingmüller, K., Steil, B., Shapiro, A. I., Witzke, V., Kostogryz, N., Gizon, L., & Klingmüller, K. (2023). Metal-rich stars are less suitable for the evolution of life on their planets. *Nature Communications, 14*, 1893.

Sheehan, M. (2007). *The international politics of space*. Routledge.

Shen, S., Jin, X., & Hao, C. (2014). Cleaning space debris with a space-based laser system. *Chinese Journal of Aeronautics, 27*(4), 805–811.

Smith, J. K. (2020). Osteoclasts and microgravity. *Life, 10*(9), 207.

Space Capital. (2022). Space Investment Quarterly 2022 Q4. https://www.spa cecapital.com/publications/space-investment-quarterly-q4-2022

Stuart, C. (2022). Could bringing back samples of Martian rock contaminate the Earth with foreign microbes? https://www.sciencefocus.com/news/ could-bringing-back-samples-of-martian-rock-contaminate-the-earth-with-for eign-microbes/

Sundahl, M. J., Broering-Jacobs, C., Johnson, C. D., Mirmina, S. A., Freeland, S., Howard, D., Sankovic, J. M., Jackson, J. W., Yormick, J. P., Kasznica, J., Hanlon, M., & Reibaldi, G. (2021). Returning to the Moon: Legal challenges as humanity begins to settle the solar system-full transcript. *The Global Business Law Review, 9*, 1.

Szocik, K. (2019). *The human factor in a mission to Mars*. Springer.

Szocik, K. (2020). Is human enhancement in space a moral duty? Missions to Mars, advanced AI and genome editing in space. *Cambridge Quarterly of Healthcare Ethics, 29*(1), 122–130.

Szocik, K., Campa, R., Rappaport, M. B., & Corbally, C. (2019). Changing the paradigm on human enhancements: The special case of modifications to counter bone loss for manned Mars missions. *Space Policy, 48*, 68–75.

Szocik, K., Wójtowicz, T., & Braddock, M. (2020). The Martian: Possible scenarios for a future human society on Mars. *Space Policy, 54*, 1–11.

Takemura, N. (2019). Astro-Green criminology: A new perspective against space capitalism outer space mining may make the same mistakes in space as we have on Earth. *Toin University of Yokohama Research Bulletin, 40*, 7–17.

Tatum, M. (2020). Drugs in space: The pharmacy orbiting the Earth. *The Pharmaceutical Journal, 305*(7939). https://doi.org/10.1211/PJ.2020.202 08033

Temmen, J. (2022). Scorched Earth: Discourses of multiplanetarity, climate change, and Martian terraforming in Finch and once upon a time i lived on Mars. *Zeitschrift Für Literaturwissenschaft Und Linguistik, 52*(3), 477–488.

Terhorst, A., & Dowling, J. A. (2022). Terrestrial analogue research to support human performance on Mars: A review and bibliographic analysis. *Space: Science & Technology* (pp. 1–18).

Tobin, J. J., Van't Hoff, M. L., Leemker, M., Van Dishoeck, E. F., Paneque-Carreño, T., Furuya, K., Harsono, D., Persson, M. V., Cleeves, L. I., Sheehan, P. D., & Cieza, L. (2023). Deuterium-enriched water ties planet-forming disks to comets and protostars. *Nature, 615*(7951), 227–230.

Tung, H. C., Bramall, N. E., & Price, P. B. (2005). Microbial origin of excess methane in glacial ice and implications for life on Mars. *Proceedings of the National Academy of Sciences, 102*(51), 18292–18296.

UNESCO. (2018). Humans are a geological force. https://en.unesco.org/cou rier/2018-2/humans-are-geological-force

United Nations Office for Outer Space Affairs. (2023). Space Law. https://www. unoosa.org/oosa/en/ourwork/spacelaw/

Van Houdt, R., Mijnendonckx, K., & Leys, N. (2012). Microbial contamination monitoring and control during human space missions. *Planetary and Space Science, 60*(1), 115–120.

Vaughan, P., & Kuś, R. (2017). From dreams to disillusionment: A socio-cultural history of the American Space Program. *Ad Americam, 18*, 75.

Vdovychenko, N. (2022). Elon Musk's SpaceX: How the 'space race' to Mars adopted The Californian Ideology. *Diggit Magazin.* https://diggitmagazine. com/articles/elon-musk-spacex

Veysi, H. (2022). Megatsunamis and microbial life on early Mars. *International Journal of Astrobiology, 21*(3), 188–196.

Walcott, R. (2014). The problem of the human: Black ontologies and "the colo- niality of our being". *Postcoloniality—decoloniality—Black critique: Joints and fissures* (pp. 93–108).

White, R. (2022). Climate change and the geographies of ecocide. In M. Bowden & A. Harkness (Eds.), *Rural transformations and rural crime* (pp. 108–124). Bristol University Press.

Williams, K. (2021). Space crime continuum: Discussing implications of the first crime in space. *Boston University International Law Journal, 39*(1), 79–108.

Witze, A. (2022). Space junk heading for Moon will add to 60+ years of lunar debris. *Nature.*

Wodecki, B. (2023). NASA turns to AI to design spacefaring hardware. https:// aibusiness.com/automation/nasa-turns-to-ai-to-design-spacefaring-hardware

Yacoubian, G. S. (2000). The (in)significance of genocidal behavior to the discipline of criminology. *Crime, Law & Social Change, 34*, 7–19.

CHAPTER 7

We Destroy, Therefore We Are: A Criminological Imagination Against the Human Species (Conclusion)

Abstract After having criminologically imagined the ways in which the human species acts as a self-formative, exploitative, and self-destructive species—and the only species living in ecological imbalance—this chapter argues that the root cause of this behaviour is the human capacity to imagine and manipulate mental images to solve problems. It is this very ability that has led to mass exploitation, extermination and the potential for self-inflicted mass extinction both on Earth and possibly in outer space in the future. Therefore, the criminological imagination of the human species concludes that it is the power of imagination itself that makes us a violent, exploitative and destructive species.

Keywords Criminological imagination · Human species · Descartes

This book has paved the way for a criminology of the human species to commence. It has done so by imaginatively exploring different aspects of our evolution as a self-formative, exploitative and self-destructive species (Dandaneau, 2021), one that lives out of continuous ecological balance and can be considered to be in bad health (McMichael, 1993). This is due in part to our intelligent survival-conducive practices (Hauptli, 1994), which have enabled us to thrive but have also had negative consequences.

© The Author(s), under exclusive license to Springer Nature   99
Switzerland AG 2023
Y. Eski, *A Criminology of the Human Species*,
Palgrave Studies in Green Criminology,
https://doi.org/10.1007/978-3-031-36092-3_7

To start the exploration of those different dimensions, Chapter 1 proposed using the criminological imagination (Young, 2011) that allows us to let go of comfortable criminological beliefs, pushing the imagination beyond conventional criminological epistemological and theoretical "truths" regarding human activities that cause crime and harm. A wide range of disciplinary perspectives have been taken into account, ranging from archaeology, astronomy, astrophysics, biology and geology, to palaeontology and planetology, in order to imagine the human species as the most powerful, exploitative and destructive species on and off Earth.

Before the criminological imagination of the human species commenced, the geological history of our planet was considered in Chapter 2. It became clear that Earth houses a violent yet regenerative system of mass extinctions and the cyclical nature of life, death, and rebirth that characterises Earth's biodiversity and ways of renewing itself. Having both terrestrial and extraterrestrial causes, as can be traced back in Earth's crust, out of such mass extinctive violence comes new life. In that sense, the current sixth mass extinction – the Anthropocene mass extinction – is not unique but is simply part of Earth's natural geohistorical cycles.

As imagined in Chapter 3, the human species is born of the mass exploitation and total annihilation of other human beings and other human-like species, such as the Neanderthal. These tendencies toward mass exploitation and total annihilation are characteristic of the human species, in particular because they result from how we are able to dehumanise our own kind through our human imagination, made possible by our precision-organ, the brain. It is all too human to dehumanise.

In Chapter 4, the human species has been imagined as an omnicidal agent. By considering ourselves to be a unique species on Earth, we (still) see ourselves as superior in relation to both the natural environment and other species. Taking no account of the balance with our natural habitat, the human species violated several ecological borders, resulting in a global-scale violent destruction, namely, the sixth mass extinction, one that is most likely irreversible. At the same time, we tend to deny the possibility of becoming extinct; that is generally not part of our imagination. It is a persistent and stubborn belief that we will overcome climate awareness by going green and embracing environment-friendly smart technologies. We are therefore a uniquely positioned species that is capable of *causing*, *accelerating* and *denying* its own, self-inflicted mass extinction (and that of other species).

In Chapter 5, the human species was imagined to be capable of conceiving itself as (having to be) a distinct being with higher-order desires; desires fulfilled through human enhancement and working toward singularity. We can imagine ourselves to literally be better versions of ourselves, by which we have managed to control and speed up human evolution. That controlled and accelerated evolution could lead to dehumanisation "upward", one in which a *homo superior* could displace and perhaps eradicate, a shared humanity. Therefore, the human species has not only caused and sped up physical mass extinction; it also has the tendency to mass-design itself out of humanity, annihilating an inner humanness. We become extinct within ourselves. It is, once more, because of our own imagination that the human species can self-inflict an inner existential change.

Finally, in Chapter 6, the human species is imagined as being capable of causing harm beyond Earth as an extraterrestrial threat. With our expanding space exploration and potential mass exploitation of celestial bodies, such as asteroids, the Moons and Mars, we run the risk of annihilating any potential extraterrestrial life and even preventing it from emerging in the first place. Our drive to control comets and asteroids as potential life-seeding objects could lead to exploitative and destructive consequences. Furthermore, our desire to expand into space requires us to redesign and enhance ourselves to survive in the harsh environments of space, potentially leading to further dehumanisation upward. These tendencies stem from our imaginative power to envision ourselves as an interplanetary species.

In sum, the criminological imagination presented throughout this book suggests that the human species exists out of ecological balance and can be considered a species living in bad health (McMichael, 1993). We are deviant in comparison to other species in the contemporary biosphere, as we are the only species on Earth that (no longer) lives in natural balance with its environment (Eldredge, 2000; Rull, 2022; Walsh, 1984). It is human action that is disrupting the natural balance between the human species and its ecology, affecting all other species on this planet (Avise et al., 2008; Eldredge, 2001; Harris et al., 2019; Janssen & Schuilenburg, 2021; Naggs, 2017; White, 2017). Furthermore, it is the human species that removes its special quality of being human through human enhancement and moving toward singularity. And it is the human species that extends its mass exploitative and totally annihilative tendencies from Earth to outer space.

More importantly, unlike other species that can imagine, like great apes, parrots and cetaceans, it is only the imagination of the human species that allows mental images to be manipulated to solve novel problems (Mitchell, 2002, 2016; Savage-Rumbaugh et al., 2009; Segerdahl et al., 2005). In having the imaginative capacities and being able to act upon them accordingly to solve problems, we have brought about mass exploitation and extermination of other human beings and other human-like species, specifically, the Neanderthal. These tendencies toward total annihilation, characteristic of the human species, result from the human imagination that dehumanises, using our precision-organ, the brain. Rooted in the very same imaginative and problem-solving capacities that have allowed us to become the dominant species on Earth, we have become a species that is capable of causing, accelerating and denying its own, self-inflicted mass extinction (and that of other species). Furthermore, our imagination and desire even for human enhancement and singularity has resulted in the potential extinction of our intangible human essence. Lastly, the destructive nature of the human species emerges in the exploitative and totally annihilative tendencies ingrained in our space expansionism, for which we are also willing to redesign ourselves. In space, we form an extraterrestrial threat to species on other planets—all because we imagine ourselves becoming an interplanetary species one day.

So, *imaginably*, it is our imagination that drives dehumanisation and forms the engine of mass exploitation, total annihilation and mass extinction on Earth. It is our imagination that pushes our potential to single out our humanness, as well as making us repeat our exploitative and totally annihilative tendencies off Earth. Given that 'human subjects are creative actors' (Karpiak, 2013: 390), it is therefore the power of imagination (Hajer, 2017) that makes us create and destroy (ourselves) on global, perhaps in the future, cosmic scales. Hence, the criminological imagination of the human species provided here that initiates a criminology of the human species is one that entails a criminological imagination *against* the human species. More specifically, it is a criminological imagination *against* the human imagination. The human species is a deviant species *because* it is able to imagine and act upon it precisely through problem solving. The power of imagination, then, next to being a most constructive human feature, is also the very cause of our (self-)destructive tendencies. It is not just the opposable thumb alone; it is also, and even more so, being able to have opposable thought that can make us a most violent, exploitative and destructive species.

In writing up this conclusion, I started to wonder whether Descartes (1637) was right after all when he said 'I think, therefore I am'. Perhaps, and put slightly differently, as the human species has the power to imagine and realise those imaginations, we could say: 'We imagine, therefore we are destruction'. An unsettling tone indeed.

References

Avise, J. C., Hubbell, S. P., & Ayala, F. J. (2008). In the light of evolution II: Biodiversity and extinction. *Proceedings of the National Academy of Sciences of the United States of America, 105*, 11453–11457.

Dandaneau, S. P. (2021). C. Wright Mills: Exact imagination, late work. *The Routledge International Handbook of C. Wright Mills Studies* (pp. 275–300). Routledge.

Descartes, R. (1637). *Discours de la Méthode. Pour bien conduire sa raison, et chercher la vérité dans les sciences*. L'Imprimerie de Ian Maire.

Eldredge, N. (2000). *Life in the balance: Humanity and the biodiversity crisis*. Princeton University Press.

Eldredge, N. (2001). The sixth extinction. https://www.biologicaldiversity.org/programs/population_and_sustainability/extinction/pdfs/Eldridge-sixth-extinction.pdf

Hajer, M. (2017). *The power of imagination*. Universiteit Utrecht.

Harris, N., Goldman, E. D., & Gibbes, S. (2019). Spatial database of planted trees (SDPT) version 1.0. *Technical Note, World Resources Institute*.

Hauptli, B. W. (1994). Rescher's unsuccessful evolutionary argument. *The British Journal for the Philosophy of Science, 45*(1), 295–301.

Janssen, J., & Schuilenburg, M. (2021). Het antropoceen. De criminologische uitdaging in de 21ste eeuw. *Tijdschrift over Cultuur & Criminaliteit, 11*(1), 3–13.

Karpiak, K. (2013). Jock Young: The criminological imagination (Book review). *Critical Criminology, 3*(21), 389–391.

McMichael, A. J. (1993). *Planetary overload: Global environmental change and the health of the human species*. Cambridge University Press.

Mitchell, R. W. (2002). *Pretending and imagination in animals and children*. Cambridge University Press.

Mitchell, R. W. (2016). Can animals imagine? In A. Kind (Ed.), *The Routledge handbook of philosophy of imagination* (pp. 326–338). Routledge.

Naggs, F. (2017). Saving living diversity in the face of the unstoppable sixth mass extinction: A call for urgent international action. *The Journal of Population and Sustainability, 1*(2), 67–81.

Rull, V. (2022). Biodiversity crisis or sixth mass extinction? Does the current anthropogenic biodiversity crisis really qualify as a mass extinction? *EMBO Reports, 23*(1), e54193.

Savage-Rumbaugh, S., Rumbaugh, D., & Fields, W. M. (2009). Empirical Kanzi: The ape language controversy revisited. *Skeptic (altadena, CA), 15*(1), 25–34.

Segerdahl, P., Fields, W., & Savage-Rumbaugh, S. (2005). *Kanzi's primal language: The cultural initiation of primates into language.* Springer.

Walsh, R. (1984). *Staying alive: The psychology of human survival.* New Science Library/Shambhala Publications.

White, R. (2017). Carbon criminals, ecocide and climate justice. In C. Holley & C. Shearing (Eds.), *Criminology and the Anthropocene* (pp. 50–80). Routledge.

Young, J. (2011). *The criminological imagination.* Polity Press.

REFERENCES

3ds.com. (2023). Sustainable life on Earth and beyond. https://www.3ds.com/insights/customer-stories/interstellar-lab-sustainable-life-on-earth

Abugre, C. (2008). Behind most mass violence lurk economic interests. In H. Melby & J. Y. Jones (Eds.), *Revisiting the heart of darkness—Explorations into genocide and other forms of mass violence* (pp. 273–280). Routledge.

Adams, P. F. (1983). A proposal for a commissioned corps of space travellers. *Akron Law Review, 17*(1), 111–130.

Adams, W. Y. (1984). The first colonial empire: Egypt in Nubia, 3200–1200 BC. *Comparative Studies in Society and History, 26*(1), 36–71.

AGU/American Geophysical Union. (2017–2019). Earth and Space Science is Essential for Society. https://agupubs.onlinelibrary.wiley.com.doi/toc/10.1002/(ISSN)2333-5084.SCISOC1

Al-Amoudi, I. (2022). Are post-human technologies dehumanizing? Human enhancement and artificial intelligence in contemporary societies. *Journal of Critical Realism, 21*(5), 516–538.

Alberro, H. (2022). HG Wells, earthly and post-terrestrial futures. *Futures, 140,* 102954.

Albertsson, T., Semenov, D., & Henning, T. (2014). Chemodynamical deuterium fractionation in the early solar nebula: The origin of water on earth and in asteroids and comets. *The Astrophysical Journal, 784*(1), 39.

Alvarez, L. W., Alvarez, W., Asaro, F., & Michel, H. V. (1980). Extraterrestrial cause for the Cretaceous-Tertiary extinction. *Science, 208*(4448), 1095–1108.

Anderson, J. R. (2020). Responses to death and dying: Primates and other mammals. *Primates, 61*(1), 1–7.

Y. Eski, *A Criminology of the Human Species*,
Palgrave Studies in Green Criminology,
https://doi.org/10.1007/978-3-031-36092-3

Anderson, J. W. (1986). NASA finds the way toward building a station fraught with legal hurdles. *Commercial Space, 2*(Spring), 59–61.

Arendt, H. (1994). On the nature of totalitarianism: An essay in understanding. In J. Kohn (Ed.), *Essays in understanding: 1930–1954* (pp. 328–360). Schocken Books.

Arendt, H. (2006). *Eichmann in Jerusalem*. Penguin.

Arendt, H. (2019). *The human condition*. The University of Chicago Press.

Aristotle. (1965). *History of animals*, Volume I: Books 1–3 (Translated by A. L. Peck). Harvard University Press.

Asgardia.space. (2023). The space nation. https://asgardia.space/

Avise, J. C., Hubbell, S. P., & Ayala, F. J. (2008). In the light of evolution II: Biodiversity and extinction. *Proceedings of the National Academy of Sciences of the United States of America, 105*, 11453–11457.

Axelrod, A. (2008). *Profiles in folly: History's worst decisions and why they went wrong*. Sterling Publishing Company Inc.

Axpe, E., Chan, D., Abegaz, M. F., Schreurs, A. S., Alwood, J. S., Globus, R. K., & Appel, E. A. (2020). A human mission to Mars: Predicting the bone mineral density loss of astronauts. *PLoS ONE, 15*(1), e0226434.

Azevedo, F. A., Carvalho, L. R., Grinberg, L. T., Farfel, J. M., Ferretti, R. E., Leite, R. E., Filho, W. J., Lent, R., & Herculano-Houzel, S. (2009, April 10). Equal numbers of neuronal and nonneuronal cells make the human brain an isometrically scaled-up primate brain. *Journal of Comparative Neurology, 513*(5), 532–541.

Baker, M. (2020). NASA Astronaut Anne McClain accused by spouse of crime in space. https://www.nytimes.com/2019/08/23/us/astronaut-space-investigation.html

Banks, W. E., d'Errico, F., Peterson, A. T., Kageyama, M., Sima, A., & Sánchez-Goñi, M. F. (2008). Neanderthal extinction by competitive exclusion. *PLoS ONE, 3*(12), 1–8.

Barakat, M. J., Field, J., & Taylor, J. (2013). The range of movement of the thumb. *The Hand, 8*, 179–182.

Barboni, M., Boehnke, P., Keller, B., Kohl, I. E., Schoene, B., Young, E. D., & McKeegan, K. D. (2017). Early formation of the Moon 4.51 billion years ago. *Science advances, 3*(1), e1602365.

Barmin, I. V., Dunham, D. W., Kulagin, V. P., Savinykh, V. P., & Tsvetkov, V. Y. (2014). Rings of debris in near-Earth space. *Solar System Research, 48*(7), 593–600.

Barnett, P. C. (2000). Reviving cyberpunk: (Re) constructing the subject and mapping cyberspace in the Wachowski Brother's film The Matrix. *Extrapolation (pre-2012), 41*(4), 359.

Barnosky, A. D., Matzke, N., Tomiya, S., Wogan, G. P., Swartz, B., Quental, T. B., Marshall, C., McGuire, J. L., Lindsey, E. L., Maguire, K. C., & Mersey,

B. (2011). Has the Earth's sixth mass extinction already arrived? *Nature, 471*(7336), 51–57.

Barton, A., Corteen, K., Scott, D., & Whyte, D. (2013). *Expanding the criminological imagination*. Routledge.

Bauman, Z. (2006). *Liquid fear*. Polity.

Bauman, Z. (2000). *Modernity and the holocaust*. Cornell University Press.

Beames, C. (2022). AI in space and its future use in warfare. https://www.for bes.com/sites/charlesbeames/2022/12/21/ai-in-space-and-its-future-use-in-warfare/amp/

Beaulieu-Laroche, L., Brown, N. J., Hansen, M., Toloza, E. H., Sharma, J., Williams, Z. M., Frosch, M. P., Cosgrove, G. R., Cash, S. S., & Harnett, M. T. (2021). Allometric rules for mammalian cortical layer 5 neuron biophysics. *Nature, 600*(7888), 274–278.

Bender, E. M. (2023). Policy makers: Please don't fall for the distractions of #AIhype. https://medium.com/@emilymenonbender/policy-makers-please-dont-fall-for-the-distractions-of-aihype-e03fa80ddbf1

Benton, M. J. (2018). Hyperthermal-driven mass extinctions: Killing models during the Permian-Triassic mass extinction. *Philosophical Transactions of the Royal Society a: Mathematical, Physical and Engineering Sciences, 376*(2130), 20170076.

Benzinga. (2022). Elon Musk says he has 'Already' uploaded his brain to the cloud. https://uk.investing.com/news/cryptocurrency-news/elon-musk-says-he-has-already-uploaded-his-brain-to-the-cloud-2689245

Best, S. (2009). Globalization of the human empire. In S. Dasgupta & J. N. Pieterse (Eds.), *Politics of globalization* (pp. 288–312). Sage.

Bezo, B., & Maggi, S. (2015). Living in "survival mode:" Intergenerational transmission of trauma from the Holodomor genocide of 1932–1933 in Ukraine. *Social Science & Medicine, 134*, 87–94.

Bianciardi, G. (2022). Is life on Mars a danger to life on Earth? NASA's Mars sample return. *Journal of Astrobiology, 11*, 14–20.

Billings, L. (2007). Ideology, advocacy, and spaceflight: Evolution of a cultural narrative. In S. J. Dick & R. D. Launius (Eds.), *Societal impact of spaceflight* (pp. 483–499). NASA.

Billings, L. (2017). Should humans colonize other planets? *Theology and Science, 15*(3), 321–332.

Bjornsdottir, R. T., & Rule, N. O. (2017). The visibility of social class from facial cues. *Journal of Personality and Social Psychology, 113*(4), 530.

Blomfield, H. D. (2004). Frontier society: Perpetuation and misrepresentation of humankind in outer space policy. https://open.library.ubc.ca/soa/cIRcle/collections/ubctheses/831/items/1.0091574

Bobylev, N. (2006). Strategic environmental assessment of urban underground infrastructure development policies. *Tunnelling and Underground Space Technology, 21*(3–4), 469.

Bohlander, M. (2021). Metalaw-What is it good for? *Acta Astronautica, 188*, 400–404.

Bostrom, N., & Roache, R. (2008). Ethical issues in human enhancement. In J. Ryberg, T. Petersen, & C. Wolf (Eds.), *New waves in applied ethics* (pp. 120–152). Palgrave Macmillan.

Bostrom, N. (2005). A history of transhumanist thought. *Journal of evolution and technology, 14*(1).

Bosworth, M., & Hoyle, C. (2012). *What is criminology?* Oxford University Press.

Boutellier, H. (2019). *A criminology of moral order.* Bristol University Press.

Bouvier, A., & Wadhwa, M. (2010). The age of the Solar System redefined by the oldest Pb–Pb age of a meteoritic inclusion. *Nature Geoscience, 3*(9), 637–641.

Brack, A. (1993). Liquid water and the origin of life. *Origins of Life and Evolution of the Biosphere, 23*(1), 3–10.

Bradley, A. M., & Wein, L. M. (2009). Space debris: Assessing risk and responsibility. *Advances in Space Research, 43*(9), 1372–1390.

Braga, A., & Logan, R. K. (2019). AI and the singularity: A fallacy or a great opportunity? *Information, 10*(2), 73.

Breitburg, D., Levin, L. A., Oschlies, A., Grégoire, M., Chavez, F. P., Conley, D. J., Garçon, V., Gilbert, D., Gutiérrez, D., Isensee, K., Jacinto, G. S. (2018). Declining oxygen in the global ocean and coastal waters. *Science, 359*(6371).

Brisman, A., & South, N. (2020). A criminology of extinction: Biodiversity, extreme consumption and the vanity of species resurrection. *European Journal of Criminology, 17*(6), 918–935.

Brisman, A., & South, N. (2013). A green-cultural criminology: An exploratory outline. *Crime Media Culture, 9*(2), 115–135.

Broderick, M. (1993). Surviving Armageddon: Beyond the imagination of disaster. *Science Fiction Studies* (pp. 362–382).

Brown, T. (2022). Can Starlink satellites be lawfully targeted? https://lieber.westpoint.edu/can-starlink-satellites-be-lawfully-targeted/

Browning, C. R. (1992). *Ordinary men: Reserve police battalion 101 and the final solution in Poland.* HarperCollins.

Bubeck, S., Chandrasekaran, V., Eldan, R., Gehrke, J., Horvitz, E., Kamar, E., Lee, P., Lee, Y. T., Li, Y., Lundberg, S., Nori, H., & Zhang, Y. (2023). Sparks of artificial general intelligence: Early experiments with gpt-4. arXiv preprint arXiv:2303.12712.

Burke, R. H. (2020). Green and species criminology. In R. H. Burke (Ed.), *Contemporary criminological theory* (pp. 300–323). Routledge.

Byers, M., Wright, E., Boley, A., & Byers, C. (2022). Unnecessary risks created by uncontrolled rocket reentries. *Nature Astronomy, 6*(9), 1093–1097.

Cairns, J., Jr. (2013). Can a species rapidly moving toward a self-inflicted extinction be considered successful? *Integrated Environmental Assessment and Management, 9*(4), 674–675.

Cameron, A. G. W. (1985). Formation and evolution of the primitive solar nebula. *Protostars and Planets I, I,* 1073.

Campbell, D. T. (1974). 'Downward causation' in hierarchically organised biological systems. *Studies in the philosophy of biology* (pp. 179–186). Palgrave.

Cannon, K. M., & Britt, D. T. (2019). Feeding one million people on Mars. *New Space, 7*(4), 245–254.

Caponi, S. (2013). Quetelet, the average man and medical knowledge. *História, Ciências, Saúde-Manguinhos, 20,* 830–847.

Carnahan, E., Vance, S. D., Cox, R., & Hesse, M. A. (2022). Surface-to-ocean exchange by the sinking of impact generated melt chambers on Europa. *Geophysical Research Letters* (p. e2022GL100287).

Castro, J. (2017). Do animals murder each other. https://www.livescience.com/60431-do-animals-murder-each-other.html

Caygill, H. (2016). Bataille and the Neanderthal extinction. *Georges Bataille and Contemporary Thought* (pp. 239–264).

Ceballos, G., Ehrlich, P. R., & Dirzo, R. (2017). Biological annihilation via the ongoing sixth mass extinction signaled by vertebrate population losses and declines. *Proceedings of the National Academy of Sciences of the United States of America, 114,* E6089–E6096.

Chatzipanagiotis, M. (2011). *The legal status of space tourists in the framework of commercial suborbital flights.* Carl Heymanns Verlag.

Chen, S. (2021). How China's space station could help power astronauts to Mars. https://www.scmp.com/news/china/science/article/3135770/how-chinas-space-station-could-help-power-astronauts-mars

Cheshire, W. P., Jr. (2015). The sum of all thoughts: Prospects of uploading the mind to a computer. *Ethics & Medicine, 31*(3), 135.

Chinn, S., Hart, P. S., & Soroka, S. (2020). Politicization and polarization in climate change news content, 1985–2017. *Science Communication, 42*(1), 112–129.

Chowdhury, R. B., Khan, A., Mahiat, T., Dutta, H., Tasmeea, T., Armaan, A. B., Fardu, F., Roy, B. B., Hossain, M. M., Khan, N. A., & Amin, A. N. (2021). Environmental externalities of the COVID-19 lockdown: Insights for sustainability planning in the Anthropocene. *Science of The Total Environment* (p. 147015).

Christensen, C., Starzyk, J., Potter, S., & Boensch, N. (2017). Start-up space: Update on investment in commercial space ventures. https://brycetech.com/downloads/Bryce_Start_Up_Space_2017.pdf

Chu, J. (2017). Scientists make huge dataset of nearby stars available to public. https://exoplanets.nasa.gov/news/1413/scientists-make-huge-dataset-of-nearby-stars-available-to-public/

Clark, M. E. (1989). Humankind at the crossroads. *Ariadne's Thread* (pp. 471–506). Palgrave Macmillan.

Clark, S. C. (2023). DARPA grant will fund hunt for drug that can keep people warm. Rice News. https://news.rice.edu/news/2023/darpa-grant-will-fund-hunt-drug-can-keep-people-warm

Clark, J., Koopmans, C., Hof, B., Knee, P., Lieshout, R., Simmonds, P., & Wokke, F. (2014). Assessing the full effects of public investment in space. *Space Policy, 30*(3), 121–134.

Clery, D. (2023). NASA unveils initial plan for multibillion-dollar telescope to find life on alien worlds. https://www.science.org/content/article/nasa-unveils-initial-plan-multibillion-dollar-telescope-find-life-alien-worlds

Cohen, S. (2013). *States of denial: Knowing about atrocities and suffering.* John Wiley & Sons.

Cohen, S. (2017). *Against criminology.* Routledge.

Colossal. (2023). The mammoth. https://colossal.com/mammoth/

Condamine, F. L., Guinot, G., Benton, M. J., & Currie, P. J. (2021). Dinosaur biodiversity declined well before the asteroid impact, influenced by ecological and environmental pressures. *Nature Communications, 12*(1), 1–16.

Condon, S. (2022). Space junk is falling from the sky. We are still not doing enough to stop it. https://www-zdnet-com.cdn.ampproject.org/c/s/www.zdnet.com/google-amp/article/space-junk-is-falling-from-the-sky-we-are-still-not-doing-enough-to-stop-it/

Cottier, C. (2021). The first 'Space hotel' plans to open in 2027. https://astronomy.com/news/2021/11/the-first-space-hotel-plans-to-open-in-2027

Cotton, D. R., Cotton, P. A., & Shipway, J. R. (2023). Chatting and cheating: Ensuring academic integrity in the era of ChatGPT. *Innovations in Education and Teaching International* (pp. 1–12).

Courtillot, V., & Gaudemer, Y. (1996). Effects of mass extinctions on biodiversity. *Nature, 381*(6578), 146–148.

Cowie, R. H., Bouchet, P., & Fontaine, B. (2022). The sixth mass extinction: Fact, fiction or speculation? *Biological Reviews, 97*(2), 640–663.

Craemer, T. (2018). Comparative analysis of reparations for the holocaust and for the transatlantic slave trade. *The Review of Black Political Economy, 45*(4), 299–324.

Creech, S., Guidi, J., & Elburn, D. (2022). Artemis: An overview of NASA's activities to return humans to the Moon. *2022 IEEE Aerospace Conference (AERO)* (pp. 1–7).

Criscuolo, F., & Sueur, C. (2020). An evolutionary point of view of animal ethics. *Frontiers in Psychology, 11,* 403.

Crist, E. (2013). Ecocide and the extinction of animal minds. In M. Bekoff (Ed.), *Ignoring nature no more: The case for compassionate conservation* (pp. 45–61). University of Chicago Press.

Crook, M., Short, D., & South, N. (2018). Ecocide, genocide, capitalism and colonialism: Consequences for indigenous peoples and glocal ecosystems environments. *Theoretical Criminology, 22*(3), 298–317.

Curry, A. (2023). Neanderthals lived in groups big enough to eat giant elephants. https://www.science.org/content/article/neanderthals-lived-gro ups-big-enough-eat-giant-elephants

Daems, T. (2006). Zygmunt Bauman en de criminologische verbeelding. *Panopticon: tijdschrift voor strafrecht, criminologie en forensisch welzijnswerk,* (6), 51–54.

Dandaneau, S. P. (2021). C. Wright Mills: Exact imagination, late work. *The Routledge International Handbook of C. Wright Mills Studies* (pp. 275–300). Routledge.

Darwin, C. (1859). *The origin of species by means of natural selection, or the preservation of favoured races in the struggle for life.* John Murray.

Darwin, C. (1872). *The expression of the emotions in man and animals.* John Murray.

Dass, R. S. (2017). The high probability of life on Mars: A brief review of the evidence. *Cosmology, 27,* 62–73.

Day, L. E., & Vandiver, M. (2000). Criminology and genocide studies: Notes on what might have been and what still could be. *Crime, Law and Social Change, 34*(1), 43–59.

De Beauvoir, S. (1952). *The second sex.* Vintage.

De Landa, M. (2000). *A thousand years of nonlinear history.* Swerve Editions.

Deans, G., & Larson, M. (2008). Growth for growth's sake: A recipe for a potential disaster. *Ivey Business Journal, 72,* 1–12.

Dekema, B. M. (2022). To infinity and beyond: Shifting the space regulatory framework to create conservation-minded expansion. *Natural Resources Journal, 62*(2), 237–256.

Demar, M. (2023). Somnium space's live forever mode made the headlines. https://somniumtimes.com/2022/04/21/somnium-spaces-live-forever-mode-made-the-headlines/

Descartes, R. (1637). *Discours de la Méthode. Pour bien conduire sa raison, et chercher la vérité dans les sciences.* L'Imprimerie de Ian Maire.

Descartes, R. (1641 [1988]). Meditations on first philosophy. In J. Cottingham, R. Stoothoff, & D. Murdoch (Eds.), *Descartes: Selected philosophical writings* (pp. 73–121). Cambridge University Press.

Deudney, D. (2020). *Dark skies: Space expansionism, planetary geopolitics, and the ends of humanity.* Oxford University Press.

Diamond, M. S., Director, H. M., Eastman, R., Possner, A., & Wood, R. (2020). Substantial cloud brightening from shipping in subtropical low clouds. *AGU Advances, 1*(1), e2019AV000111.

Diamond, J. (2013). *The rise and fall of the third chimpanzee*. Random House.

Dickens, P., & Ormrod, J. (2008). Who really won the space race? *Monthly Review, 59*(9), 30–37.

Diederiks-Verschoor, I. (1967). Some observations regarding the treaty on space law. *Proceedings on the Law of Outer Space, 10*, 164–169.

Diederiks-Verschoor, I. (1979). Space law as it affects domestic law. *Journal of Space Law, 7*(1), 39–46.

Dobynde, M. I., Shprits, Y. Y., Drozdov, A. Y., Hoffman, J., & Li, J. (2021). Beating 1 sievert: Optimal radiation shielding of astronauts on a mission to Mars. *Space Weather, 19*(9), e2021SW002749.

Donati, P. (2010). *Relational sociology: A new paradigm for the social sciences*. Cambridge University Press.

Dongarwar, D. (2022). Extreme acts of violence: Infanticide and associated social constructs. *Handbook of anger, aggression, and violence* (pp. 1–17). Springer International Publishing.

Donoghue, D. (1979). Review of Hannah Arendt's *The Life of the Mind. The Hudson Review, 32*(2), 281–288.

Douglas, A. (2022). 10 animals that kill the most humans. https://www.worlda tlas.com/animals/10-animals-that-kill-the-most-humans.html

Doyle, B. (2015). The Postapocalyptic Imagination. *Thesis Eleven, 131*(1), 99–113.

Drescher, S. (2018). The Atlantic slave trade and the holocaust: A comparative analysis. In A. S. Rosenbaum (Ed.), *Is the Holocaust unique?: Perspectives on comparative genocide* (pp. 103–124). Routledge.

Drolet, M. J., Désormeaux-Moreau, M., Soubeyran, M., & Thiébaut, S. (2020). Intergenerational occupational justice: Ethically reflecting on climate crisis. *Journal of Occupational Science, 27*(3), 417–431.

Dunnett, O., Maclaren, A., Klinger, J., Maria, K., Lane, D., & Sage, D. (2019). Geographies of outer space: Progress and new opportunities. *Progress in Human Geography, 43*(2), 314–336.

Durrani, H. A. (2018). Interpreting space resources obtained: Historical and postcolonial interventions in the law of commercial space mining. *Columbia Journal of Transnational Law, 57*, 403–460.

Earl, D. S. (2011). The Joshua delusion: Rethinking genocide in the Bible. *The Joshua Delusion* (pp. 1–190).

Edgar, A., & Sedgewick, P. (1999). *Cultural theory: The key concepts*. Routlege.

Edwards, C. M., & Cichan, T. (2021). From the Moon to Mars base camp: An updated architecture that builds on artemis. *ASCEND 2021* (p. 4137).

Ehrmann, M. (1948). *The desiderata of love: A collection of poems for the beloved.* Crown.

Ehrmann, M., & Ehrmann, B. (1948). *The poems of Max Ehrmann.* Bruce Humphries.

Eiseley, L. (1958). *Darwin's century.* Doubleday.

Eldredge, N. (2000). *Life in the balance: Humanity and the biodiversity crisis.* Princeton University Press.

Eldredge, N. (2001). The sixth extinction. https://www.biologicaldiversity.org/programs/population_and_sustainability/extinction/pdfs/Eldridge-sixth-extinction.pdf

Elliott, A. (2019). *The culture of AI: Everyday life and the digital revolution.* Routledge.

Elms, A. K., Tremblay, P. E., Gänsicke, B. T., Koester, D., Hollands, M. A., Gentile Fusillo, N. P., Cunningham, T., & Apps, K. (2022). Spectral analysis of ultra-cool white dwarfs polluted by planetary debris. *Monthly Notices of the Royal Astronomical Society, 517*(3), 4557–4574.

Emery, A. J. (2013). *The case for spacecrime: The rise of crime and piracy in the space domain.* https://apps.dtic.Mil/sti/pdfs/ADA615000.pdf

Erwin, S., & Werner, D. (2022). Dark clouds, silver linings: Five ways war in Ukraine is transforming the space domain. https://spacenews.com/dark-clouds-silver-linings-five-ways-war-in-ukraine-is-transforming-the-space-domain/

ESA. (2022). Space and daily life. https://www.esa.int/Science_Exploration/Human_and_Robotic_Exploration/Education/Space_and_daily_life

ESA. (2023). Guide morphing rovers across alien world in evolutionary computing contest. https://www.esa.int/Enabling_Support/Space_Engineering_Technology/Guide_morphing_rovers_across_alien_world_in_evolutionary_computing_contest

Eski, Y. (2011). 'Port of call': Toward a criminology of port security. *Criminology & Criminal Justice, 11*(5), 415–431.

Eski, Y. (2020). An existentialist victimology of genocide? In Y. Eski (Ed.), *Genocide and victimology* (pp. 6–22). Routledge.

Eski, Y. (2022a). Omnia cadunt. Naar een victimologische verbeelding van onze vergankelijkheid Omnia cadunt (Translated from Toward a victimological imagination of our transience). *Tijdschrift over Cultuur & Criminaliteit, 12*(1), 58–71.

Eski, Y. (2022b). *A criminological biography of an arms dealer.* Routledge.

Eski, Y. (2023). Crime among the stars: Space criminology and the domain of limitlessness (in Dutch: Criminaliteit Tussen de Sterren: Ruimtecriminologie en het Domein van het Grenzeloze). *De Criminoloog, 28,* 7. https://research.vu.nl/files/233723169/Nieuwsbrief28_def.pdf

Eski, Y., & Walklate, S. (2020). A victimological imagination of genocide. In Y. Eski (Ed.), *Genocide and victimology* (pp. 202–210). Routledge.

European Commission. (2023). An EU Space Strategy for Security and Defence to ensure a stronger and more resilient EU. https://ec.europa.eu/commission/presscorner/detail/en/ip_23_1601

Ferrell, J., Hayward, K., Morrison, W., & Presdee, M. (2004). *Cultural criminology unleashed*. Routledge.

Ferrell, J., & Sanders, C. (1995). *Cultural criminology: An invitation*. UPNE.

Ferris, I. (2000). *Enemies of Rome: Barbarians through Roman eyes*. Sutton Publishing.

Foust, J. (2023). White House proposes $27.2 billion for NASA in 2024. https://spacenews.com/white-house-proposes-27-2-billion-for-nasa-in-2024/

Francis, M. R. (2017). The sixth mass extinction: We aren't the dinosaurs, we're the asteroid. https://www.thedailybeast.com/the-sixth-mass-extinction-we-arent-the-dinosaurs-were-the-asteroid

Friedrichs, D. O. (1994). Crime wars and peacemaking criminology. *Peace Review, 6*(2), 159–164.

Froehlich, A. (2021). *Assessing a Mars agreement including human settlements*. Springer.

Gabriel, T. S., Hardgrove, C., Achilles, C. N., Rampe, E. B., Rapin, W. N., Nowicki, S., Czarnecki, S., Thompson, L., Nikiforov, S., Litvak, M., Mitrofanov, I., & McAdam, A. (2022). On an extensive late hydrologic event in Gale Crater as indicated by water-rich fracture halos. *Journal of Geophysical Research: Planets, 127*(12), e2020JE006600.

Galey, P. (2022). 2022 Was Europe's hottest summer on record by a 'Substantial margin'. https://www.sciencealert.com/2022-was-europes-hottest-summer-on-record-by-a-substantial-margin

Gallagher, W. (2014). Platonic and attic laws on slavery. *The Compass, 1*(1), 1–4.

Galway-Witham, J., Cole, J., & Stringer, C. (2019). Aspects of human physical and behavioural evolution during the last 1 million years. *Journal of Quaternary Science, 34*(6), 355–378.

Gambacurta, A., Merlini, G., Ruggiero, C., Diedenhofen, G., Battista, N., Bari, M., Balsamo, M., Piccirillo, S., Valentini, G., Mascetti, G., & Maccarrone, M. (2019). Human osteogenic differentiation in Space: Proteomic and epigenetic clues to better understand osteoporosis. *Scientific Reports, 9*(1), 8343.

Garland, D. (2012). *The culture of control: Crime and social order in contemporary society*. University of Chicago Press.

Garland, D. (1992). Criminological knowledge and its relation to power. Foucault's genealogy and criminology today. *British Journal of Criminology, 32*, 403–422.

Garreaud, R. D., Clem, K., & Veloso, J. V. (2021). The South Pacific Pressure Trend Dipole and the Southern Blob. *Journal of Climate, 34*(18), 7661–7676.

Gates, B. (2014). The deadliest animal in the world. *Mosquito Week. The Gates Notes LLC.*

Gatti, L. V., Basso, L. S., Miller, J. B., Gloor, M., Gatti Domingues, L., Cassol, H. L., Tejada, G., Aragão, L. E., Nobre, C., Peters, W., & Marani, L. (2021). Amazonia as a carbon source linked to deforestation and climate change. *Nature, 595*(7867), 388–393.

Gaudzinski-Windheuser, S., Kindler, L., MacDonald, K., & Roebroeks, W. (2023). Hunting and processing of straight-tusked elephants 125.000 years ago: Implications for Neanderthal behavior. *Science Advances, 9*(5), eadd8186.

Gee, H. (2021). Humans are doomed to go extinct. https://www.scientificam erican.com/article/humans-are-doomed-to-go-extinct/

Giubilini, A., & Sanyal, S. (2015). The ethics of human enhancement. *Philosophy Compass, 10*(4), 233–243.

Glikson, A. Y. (2021). *The fatal species: From warlike primates to planetary mass extinction.* Springer.

Gohd, C. (2020). Astronaut Anne McClain's estranged wife charged with lying about alleged 'space crime.' https://space.com/astronaut-anne-mcclain-wife-charged-lying-space-crime.html

Golding, W. (1955). *The inheritors.* Faber & Faber.

Gómez, J. M., Verdú, M., González-Megías, A., & Méndez, M. (2016). The phylogenetic roots of human lethal violence. *Nature, 538*(7624), 233–237.

Gorove, S. (1972a). Criminal jurisdiction in outer space. *The International Lawyer, 6*(2), 313–323.

Gorove, S. (1972b). Pollution and outer space: A legal analysis and appraisal. *New York University Journal of International Law and Politics, 5*(1), 53–66.

Government of the Netherlands. (2022). Speech by Prime Minister Mark Rutte about the role of the Netherlands in the history of slavery. https://www.gov ernment.nl/documents/speeches/2022/12/19/speech-by-prime-minister-mark-rutte-about-the-role-of-the-netherlands-in-the-history-of-slavery

Grabosky, P. N., & Smith, R. G. (1998). *Crime in the digital age: Controlling telecommunications and cyberspace illegalities.* Transaction Publishers.

Graham, J. R. (1996). Dinantian river systems and coastal zone sedimentation in northwest Ireland. *Geological Society, London, Special Publications, 107*(1), 183–206.

Greaves, J. S., Richards, A., Bains, W., Rimmer, P. B., Sagawa, H., Clements, D. L., Seager, S., Petkowski, J. J., Sousa-Silva, C., Ranjan, S., Drabek-Maunder, E., & Hoge, J. (2021). Phosphine gas in the cloud decks of Venus. *Nature Astronomy, 5*(7), 655–664.

Greely, H. T. (2019). CRISPR'd babies: Human germline genome editing in the 'He Jiankui affair.' *Journal of Law and the Biosciences, 6*(1), 111–183.

Green, P. J., & Ward, T. (2000). State crime, human rights, and the limits of criminology. *Social Justice, 27*(1), 101–115.

Greenbaum, D. (2020). Space debris puts exploration at risk. *Science, 370*(6519), 922–922.

Greenbaum, G., Getz, W. M., Rosenberg, N. A., Feldman, M. W., Hovers, E., & Kolodny, O. (2019). Disease transmission and introgression can explain the long-lasting contact zone of modern humans and Neanderthals. *Nature Communications, 10*(1), 5003.

Greenfieldboyce, N. (2022). NASA is bringing rocks back from Mars, but what if those samples contain alien life? https://www.npr.org/2022/05/04/109 5645081/nasa-is-bringing-rocks-back-from-mars-but-what-if-those-samples-contain-alien-li

Gregg, J. (2021). Space Biz. *The cosmos economy* (pp. 135–144). Copernicus.

Greshko, M. (2019). Wat waren de 5 massa-uitstervings en wat veroorzaakte ze? https://www.nationalgeographic.nl/wetenschap/2019/09/wat-waren-de-5-massa-uitstervings-en-wat-veroorzaakte-ze

Grush, L. (2023). Asteroid-mining startup AstroForge to launch first space missions this year. https://www.bloomberg.com/news/articles/2023-01-24/asteroid-mining-startup-astroforge-plans-first-platinum-refining-space-missions

Gunasekara, S. G. (2010). The march of science: Fourth amendment implications on remote sensing in criminal law. *Journal of Space Law, 36*(1), 115–142.

Guston, D. H. (2010). Human enhancement. *Sage reference encyclopedia of nanoscience and society* (pp. 1–6). Sage.

Hagan, J., & Rymond-Richmond, W. (2008). The collective dynamics of racial dehumanization and genocidal victimization in Darfur. *American Sociological Review, 73*(6), 875–902.

Hajer, M. (2017). *The power of imagination.* Universiteit Utrecht.

Harrington, C., & Shearing, C. (2016). *Security in the Anthropocene: Reflections on safety and care.* Transcript Verlag.

Harris, N., Goldman, E. D., & Gibbes, S. (2019). Spatial database of planted trees (SDPT) version 1.0. *Technical Note, World Resources Institute.*

Hartzok, A. (2001). Democracy, earth rights, and next economy. Paper presented at the 21st annual E. F. Schumacher lectures. http://www.earthrights.net/docs/schumacher.html

Hashimoto, T., Horikawa, D. D., Saito, Y., Kuwahara, H., Kozuka-Hata, H., Shin-i, T., Minakuchi, Y., Ohishi, K., Motoyama, Y., Aizu, T., & Kunieda, T. (2016). Extremotolerant tardigrade genome and improved radiotolerance of human cultured cells by tardigrade-unique protein. *Nature Communications, 7*(1), 12808.

Haslam, N. (2019). The many roles of dehumanization in genocide. In L. S. Newman (Ed.), *Confronting humanity at its worst: Social psychological perspectives on genocide* (pp. 199–139). OUP.

Hattenstone, S. (2023). Tech guru Jaron Lanier: 'The danger isn't that AI destroys us. It's that it drives us insane.' https://www.theguardian.com/technology/2023/mar/23/tech-guru-jaron-lanier-the-danger-isnt-that-ai-destroys-us-its-that-it-drives-us-insane

Haughney, E. W. (1963, June 18–20). Criminal responsibility in outer space. Paper presented at the Conference on Space Science and Space Law, University of Oklahoma, Norman, Oklahoma.

Hauptli, B. W. (1994). Rescher's unsuccessful evolutionary argument. *The British Journal for the Philosophy of Science, 45*(1), 295–301.

Hauskeller, M. (2012). My brain, my mind, and I: Some philosophical assumptions of mind-uploading. *International Journal of Machine Consciousness, 4*(01), 187–200.

Hawkesworth, C. J., Dhuime, B., Pietranik, A. B., Cawood, P. A., Kemp, A. I., & Storey, C. D. (2010). The generation and evolution of the continental crust. *Journal of the Geological Society, 167*(2), 229–248.

Haynes, G. (2018). The evidence for human agency in the late pleistocene megafaunal extinctions. In D. A. DellaSala & M. I. Goldstein (Eds.), *The Encyclopedia of the Anthropocene* (pp. 219–226). Elsevier.

Hayward, K. (2016). Space–the final frontier: Criminology, the city and the spatial dynamics of exclusion. *Cultural criminology unleashed* (pp. 169–180). Routledge.

He, H., Ji, J., Zhang, Y., Hu, S., Lin, Y., Hui, H., Hao, J., Li, R., Yang, W., Tian, H., Zhang, C., & Wu, F. (2023). A solar wind-derived water reservoir on the Moon hosted by impact glass beads. *Nature Geoscience* (pp. 1–7).

Hediger, R. (2018). Animal suicide and "anthropodenial." *Animal Sentience, 20*(16), 1–3.

Hein, A. M., Saidani, M., & Tollu, H. (2018). Exploring potential environmental benefits of asteroid mining. ArXiv:1810.04749.

Heinämaa, S. (2018). Strange vegetation: Emotional undercurrents of Tove Jansson's Moominvalley in November. *SATS, 19*(1), 41–67.

Heise, U. K. (2016). *Imagining extinction.* University of Chicago Press.

Hellmann, K. U., & Luedicke, M. K. (2018). The throwaway society: A look in the back mirror. *Journal of Consumer Policy, 41*, 83–87.

Hellweg, C. E., & Baumstark-Khan, C. (2007). Getting ready for the manned mission to Mars: The astronauts' risk from space radiation. *Naturwissenschaften, 94*(7), 517–526.

Hesselbo, S. P., Robinson, S. A., Surlyk, F., & Piasecki, S. (2002). Terrestrial and marine extinction at the Triassic-Jurassic boundary synchronized with major carbon-cycle perturbation: A link to initiation of massive volcanism? *Geology*, *30*(3), 251–254.

Hird, M. J. (2002). Re(pro)ducing sexual difference. *Parallax*, *8*(4), 94–107.

Hirschfeld, A. R., & Blackmer, S. (2021). Beyond acedia and wrath: Life during the climate apocalypse. *Anglican Theological Review*, *103*(2), 196–207.

Hoelzle, M., Darms, G., Lüthi, M. P., & Suter, S. (2011). Evidence of accelerated englacial warming in the Monte Rosa area, Switzerland/Italy. *The Cryosphere*, *5*(1), 231–243.

Hoffman, J. A., Hecht, M. H., Rapp, D., Hartvigsen, J. J., SooHoo, J. G., Aboobaker, A. M., McClean, J. B., Liu, A. M., Hinterman, E. D., Nasr, M., & Hariharan, S. (2022). Mars Oxygen ISRU Experiment (MOXIE)—Preparing for human Mars exploration. *Science Advances*, *8*(35), 1–6.

Hofstadter, D. R. (1995). *Fluid concepts and creative analogies: Computer models of the fundamental mechanisms of thought*. Basic books.

Horkheimer, M., & Adorno, T. W. (1997). *Dialectic of enlightenment*. Verso.

Hornsey, M. J., Fielding, K. S., Harris, E. A., Bain, P. G., Grice, T., & Chapman, C. M. (2022). Protecting the planet or destroying the universe? Understanding reactions to space mining. *Sustainability*, *14*(7), 4119.

Houser, K. (2023). This incredibly life-like robot hand can be made for just $2,800. https://www.freethink.com/robots-ai/humanoid-robots-clone-hand

Howell, A. (2015). Resilience, war, and austerity: The ethics of military human enhancement and the politics of data. *Security Dialogue*, *46*(1), 15–31.

Hublin, J. J. (2012). The earliest modern human colonization of Europe. *Proceedings of the National Academy of Sciences*, *109*(34), 13471–13472.

Huebert, J. H., & Block, W. (2006). Space Environmentalism, Property Rights, and the Law. *University of Memphis Law Review*, *37*, 281.

Hume, D. (1740 [1978]). A Treatise of Human Nature, L. A. Selby-Bigge & P. H. Nidditch (Eds.), 2nd edition. OUP.

Hunt, P. (2017). *Ancient Greek and Roman slavery*. John Wiley & Sons.

Huxley, A. (1932). *Brave new world*. Chatto & Windus.

Ikemoto, L. C. (2017). DIY Bio: Hacking life in biotech's backyard. *University of California, Davis, Law Review*, *51*, 539.

Ilardo, M. A., Moltke I., Korneliussen, T. S., Cheng, J., Stern, A. J., Racimo, F., de Barros Damgaard, P., Sikora, M., Seguin-Orlando, A., Rasmussen, S., & van den Munckhof, I. C. (2018). Physiological and genetic adaptations to diving in sea nomads. *Cell*, *173*(3), 569–580.

IPCC. (2021). Climate change 2021: The physical science basis report. https://www.ipcc.ch/report/ar6/wg1/downloads/report/IPCC_AR6_WGI_SPM.pdf

IPCC. (2022). Climate change 2022: Impacts, adaptation and vulnerability. https://www.ipcc.ch/report/ar6/wg2/downloads/report/IPCC_AR6_WGII_FullReport.pdf

Jackson, T. (2022). Billionaire space race: The ultimate symbol of capitalism's obsession with growth. https://nextnature.net/magazine/story/2022/billio naire-space-race-the-ultimate-symbol-of-capitalism-s-obsession-with-growth

Jamieson, D. (1983). Killing persons and other beings. *Ethics and animals* (pp. 135–146).

Janssen, J., & Schuilenburg, M. (2021). Het antropoceen. De criminologische uitdaging in de 21ste eeuw. *Tijdschrift over Cultuur & Criminaliteit, 11*(1), 3–13.

Johnson, M. R. (2020). Mining the high frontier: Sovereignty, property and humankind's common heritage in outer space (Doctoral dissertation). http://hdl.handle.net/10453/142380

Johnson, N., & Klinkrad, H. (2009). The International space station and the space debris environment: 10 Years on. *5th European Conference on Space Debris* (No. JSC-CN-17944). https://ntrs.nasa.gov/citations/20090004997

Jones, A. (2016). *Genocide: A comprehensive introduction.* Routledge.

Joseph, R. G., Gibson, C., Wolowski, K., Bianciardi, G., Kidron, G. J., Armstrong, R. A., & Schild, R. (2022). Evolution of life in the oceans of Mars? Episodes of global warming, flooding, rivers, lakes, and chaotic orbital obliquity. *Journal of Astrobiology, 13*(2022), 14–126.

Joyce, C. (2020). Responses to apocalypse: Early Christianity and extinction rebellion. *Religions, 11*(8), 384.

Kaag, S. (2021). LinkedIn post over klimaatverandering. https://www.linkedin. com/posts/sigrid-kaag_ipcc-ongepercentC3percentABvenaarde-klimaatveran dering-leidt-activity-6830433144768909312-UIs

Kaiser, R. I., Stockton, A. M., Kim, Y. S., Jensen, E. C., & Mathies, R. A. (2013). On the formation of dipeptides in interstellar model ices. *The Astrophysical Journal, 765*(2), 111.

Kaku, M. (2018). *The future of humanity: Terraforming Mars, interstellar travel, immortality, and our destiny beyond.* Penguin.

Kaku, M. (2023). Michio Kaku: 3 mind-blowing predictions about the future. https://bigthink.com/the-well/predictions-for-the-future/

Kamien, R. (2008). *Music: An appreciation, sixth brief edition* (student). McGraw-Hill Higher Education.

Kant, I. (1781[1999]). *Critique of Pure Reason*, P. Guyer, & A. W. Wood (Eds.). Cambridge University Press.

Kaplan, M. (2009). Survey of space debris reduction methods. *In AIAA space 2009 Conference & Exposition* (p. 6619).

Kargel, J. S., & Lewis, J. S. (1993). The composition and early evolution of Earth. *Icarus, 105*(1), 1–25.

Karpiak, K. (2013). Jock Young: The criminological imagination (Book review). *Critical Criminology, 3*(21), 389–391.

Kass, L. R. (2003). Ageless bodies, happy souls: Biotechnology and the pursuit of perfection. *The New Atlantis, 1*, 9–28.

Katz, F. E. (1993). *Ordinary people and extraordinary evil; a Report on the beguilings of evil*. State University of New York Press.

Kaufmann, P., Kuch, H., Neuhaeuser, C., & Webster, E. (2010). *Humiliation, degradation, dehumanization: Human dignity violated*. Springer.

Keller, J. (2021). The inside story behind the Pentagon's ill-fated quest for a real-life 'Iron Man' suit. https://taskandpurpose.com/news/pentagon-powered-armor-iron-man-suit/

Kern, S. (2021). No, billionaires won't "escape" to space while the world burns. https://salon.com/2021/07/07/no-billionaires-wont-escape-to-space-while-the-world-burns/

Kerns, J. (2017). Mining materials in outer space. *Machine Design, 89*(5), 9–10.

Kerr, P. (2002). Saved from extinction: Evolutionary theorising, politics and the state. *The British Journal of Politics and International Relations, 4*(2), 330–358.

Khan, C. (2022). Will we ever see pictures of the big bang? We ask an expert. https://www.theguardian.com/lifeandstyle/2022/sep/23/will-we-ever-see-pictures-of-the-big-bang-we-ask-an-expert

Kiernan, B. (2008). *Blood and soil: A world history of genocide and extermination from Sparta to Darfur*. Yale University Press.

Kiernan, B. (2004). The first genocide: Carthage, 146 BC. *Diogenes, 51*(3), 27–39.

Kiessling, W., & Aberhan, M. (2007). Geographical distribution and extinction risk: Lessons from Triassic-Jurassic marine benthic organisms. *Journal of Biogeography, 34*(9), 1473–1489.

Kilic, C. (2022). Mars is littered with 15,694 pounds of human trash from 50 years of robotic exploration. https://theconversation.com/mars-is-littered-with-15-694-pounds-of-human-trash-from-50-years-of-robotic-exploration-188881

King, B. J. (2016). Animal mourning: Précis of How animals grieve. *Animal Sentience, 4*(1), 1–6.

Knight, A. (2017). This woman spent a year 'on Mars' as a psychological experiment. https://www.vice.com/en/article/zmb4q3/this-woman-spent-a-year-on-mars-as-a-psychological-experiment

Knoppers, B. M., & Joly, Y. (2007). Our social genome? *Trends in Biotechnology, 25*(7), 284–288.

Kochi, T., & Ordan, N. (2008). An argument for the global suicide of humanity. *Borderlands, 7*(3), 1–21.

Koene, R. A. (2014). Feasible mind uploading. *Intelligence unbound: Future of uploaded and machine minds* (pp. 90–101).

Koh, L., et al. (2004). Species coextinctions and the biodiversity crisis. *Science*, *305*(5690), 1632–1634.

Kolbert, E. (2014). *The sixth extinction: An unnatural history*. A&C Black.

Kotzé, J., & Antonopoulos, G. A. (2022). Con Air: Exploring the trade in counterfeit and unapproved aircraft parts. *The British Journal of Criminology*, 1–16.

Krauss, L. M. (2007). *The physics of Star Trek*. Basic Books.

Kuipers, A. (2022, September). Galactic entrepreneurship. *Holland Herald* (p. 22).

Kumar, A. (2012). The pollex-index complex and the kinetics of opposition. *Journal of Health and Allied Sciences NU, 2*(04), 80–87.

Kurzweil, R., Benek, C., Boss, J., Reed-Butler, P., Caligiuri, M., Dabrowski, I. J., Graves, M., Haynor, A. L., Molhoek, B., Robinson, P., Stroda, U., & Weissenbacher, A. (2020). *Spiritualities, ethics, and implications of human enhancement and artificial intelligence*. Vernon Press.

Lampkin, J. (2020). Mapping the terrain of an astro-green criminology: A case for extending the green criminological lens outside of planet Earth. *Astropolitics, 18*(3), 238–259.

Lampkin, J. (2021). Should criminologists be concerned with outer space?: A proposal for an 'Astro-criminology'. https://research.leedstrinity.ac.uk/en/publications/should-criminologists-be-concerned-with-outer-space-a-propos al-fo

Lampkin, J., & White, R. (2023). *Space criminology: Analysing human relationships with outer space*. Springer/Palgrave Macmillan.

Lampkin, J., & Wyatt, T. (2022). Widening the scope of "Earth" jurisprudence and "Green" criminology? Toward preserving extra-terrestrial heritage sites on celestial bodies. In J. Gacek & R. Jochelson (Eds.), *Green criminology and the law* (pp. 309–329). Springer.

Lang, J. (2017). Explaining genocide: Hannah Arendt and the social-scientific concept of dehumanization. *The Anthem Companion to Hannah Arendt, 187*, 188.

Lasch, C. (1979). *The culture of narcissism*. Norton.

Leblanc, A., Matsumoto, T., Jones, J., Shapiro, J., Lang, T., Shackelford, L., & Ohshima, H. (2013). Bisphosphonates as a supplement to exercise to protect bone during long-duration spaceflight. *Osteoporosis International, 24*, 2105–2114.

Lee, P., Shubham, S., & Schutt, J. W. (2023). A relict glacier near Mars' equator: Evidence for recent glaciation and volcanism in Eastern Noctis Labyrinthus. *54th Lunar and Planetary Science Conference 2023* (LPI Contrib. No. 2806). https://www.hou.usra.edu/meetings/lpsc2023/pdf/2998.pdf

Lees, A. C., Attwood, S., Barlow, J., & Phalan, B. (2020). Biodiversity scientists must fight the creeping rise of extinction denial. *Nature Ecology & Evolution*, 4(11), 1440–1443.

Lemaître, G. (1931 [2013]). A homogeneous universe of constant mass and increasing radius accounting for the radial velocity of extra-galactic nebulae. *A source book in astronomy and astrophysics, 1900–1975*. Harvard University Press.

Leslie, J. (2002). *The end of the world: The science and ethics of human extinction.* Routledge.

Levchenko, I., Xu, S., Mazouffre, S., Keidar, M., & Bazaka, K. (2021). Mars colonization: Beyond getting there. *Terraforming Mars* (pp. 73–98).

Li, J., Bonkowski, M. S., Moniot, S., Zhang, D., Hubbard, B. P., Ling, A. J., Rajman, L. A., Qin, B., Lou, Z., Gorbunova, V., Aravind, L., & Sinclair, D. A. (2017). A conserved NAD+ binding pocket that regulates protein-protein interactions during aging. *Science, 355*(6331), 1312–1317.

Li, M., Grasby, S. E., Wang, S. J., Zhang, X., Wasylenki, L. E., Xu, Y., ... & Shen, Y. (2021). Nickel isotopes link Siberian Traps aerosol particles to the end-Permian mass extinction. *Nature Communications, 12*(1), 2024.

Lin, P., Bekey, G., & Abney, K. (2008). *Autonomous military robotics: Risk, ethics, and design.* California Polytechnic State University.

Lockwood, G. M. (2011). Social egg freezing: The prospect of reproductive 'immortality' or a dangerous delusion? *Reproductive Biomedicine Online, 23*(3), 334–340.

Loder, R. E. (2018). Asteroid mining: Ecological jurisprudence beyond Earth. *Virginia Environmental Law Journal, 36*(3), 275–317.

Logan, R. K. (2018). The Anthropocene and climate change: An existential crisis. https://www.researchgate.net/profile/Robert_Logan5/publication/322437374_The_Anthropocene_and_Climate_Change_An_Existential_Cri sis/links/5a58cb73aca2727d60814d66/The-Anthropocene-and-Climate-Change-An-Existential-Crisis

Logsdon, J. M. (2015). Why did the United States retreat from the moon? *Space Policy, 32*, 1–5.

Loizou, J. (2006). Turning space tourism into commercial reality. *Space Policy, 22*(4), 289–290.

Lombard, M., & Phillipson, L. (2010). Indications of bow and stone-tipped arrow use 64,000 years ago in KwaZulu-Natal, South Africa. *Antiquity, 84*(325), 635–648.

Lombroso, C. (1876). *L'uomo delinquent. Studiato in rapporto alla antropologia, alla medicina legale ed alle discipline carcerarie.* Ulrico Hoepli, Libraio-editore.

Loughran, J. (2016). Micro spaceships powered by lasers to search for alien life [News Briefing]. *Engineering & Technology, 11*(4), 18–18.

Lovens, P. (2023). Sans ces conversations avec le chatbot Eliza, mon mari serait toujours là. https://www.lalibre.be/belgique/societe/2023/03/28/sans-ces-conversations-avec-le-chatbot-eliza-mon-mari-serait-toujours-la-LVSLWP C5WRDX7J2RCHNWPDST24/

Luo, M., Shi, G. R., Buatois, L. A., & Chen, Z. Q. (2020). Trace fossils as proxy for biotic recovery after the end-Permian mass extinction: A critical review. *Earth-Science Reviews, 203*, 103059.

Lynteris, C. (2020). *Human extinction and the pandemic imaginary*. Taylor & Francis.

Lyson, T. R., Miller, I. M., Bercovici, A. D., Weissenburger, K., Fuentes, A. J., Clyde, W. C., Hgadorn, J. W., Butrim, M. J., Johnson, K. R., Fleming, R. F., & Barclay, R. S. (2019). Exceptional continental record of biotic recovery after the Cretaceous-Paleogene mass extinction. *Science, 366*(6468), 977–983.

Maccarini, A. M. (2021). The social meanings of perfection: Human self-understanding in a post-human society. *What is essential to being human?* (pp. 197–213). Routledge.

Macip, C. C., Hasan, R., Hoznek, V., Kim, J., Metzger IV, L. E., Sethna, S., & Davidsohn, N. (2023). Gene therapy mediated partial reprogramming extends lifespan and reverses age-related changes in aged mice. *BioRxiv, 1.*

Madani, A., Krause, B., Greene, E. R., Subramanian, S., Mohr, B. P., Holton, J. M., Olmos, Jr., J. L., Xiong, C., Sun, Z. Z., Socher, R., Fraser, J. S., & Naik, N. (2023). Large language models generate functional protein sequences across diverse families. *Nature Biotechnology* (pp. 1–8).

Magnuson, E., Altshuler, I., Fernández-Martínez, M. Á., Chen, Y. J., Maggiori, C., Goordial, J., & Whyte, L. G. (2022). Active lithoautotrophic and methane-oxidizing microbial community in an anoxic, sub-zero, and hyper-saline High Arctic spring. *The ISME Journal* (pp. 1–11).

Maier-Katkin, D., Mears, D. P., & Bernard, T. J. (2009). Toward a criminology of crimes against humanity. *Theoretical Criminology, 13*(2), 227–255.

Malasse, A. D., Debénath, A., & Pelegrin, J. (1992). On new models for the Neanderthal debate. *Current Anthropology, 33*(1), 49–54.

Malhan, K., Ibata, R. A., Sharma, S., Famaey, B., Bellazzini, M., Carlberg, R. G., & Thomas, G. F. (2022). The global dynamical atlas of the Milky Way mergers: Constraints from Gaia EDR3–based orbits of globular clusters, stellar streams, and satellite galaxies. *The Astrophysical Journal, 926*(2), 107.

Marikar, S. (2018). The rich are planning to leave this wretched planet. https:///nytimes.com/2018/06/09/style/axiom-space-travel.html

Mark, C. P., & Kamath, S. (2019). Review of active space debris removal methods. *Space Policy, 47*, 194–206.

Martin, E. S., Whitten, J. L., Kattenhorn, S. A., Collins, G. C., Southworth, B. S., Wiser, L. S., & Prindle, S. (2023). Measurements of regolith thicknesses on Enceladus: Uncovering the record of plume activity. *Icarus, 392*, 115369.

Mattei, U., & Nader, L. (2008). *Plunder: When the rule of law is illegal*. John Wiley & Sons.

Maynard, A. (2022). 'Jurassic World' scientists still haven't learned that just because you can doesn't mean you should – real-world genetic engineers can learn from the cautionary tale. https://theconversation.com/jurassic-world-scientists-still-havent-learned-that-just-because-you-can-doesnt-mean-you-should-real-world-genetic-engineers-can-learn-from-the-cautionary-tale-184369

Mayor, A. (2019). *Gods and robots: Myths, machines, and ancient dreams of technology*. Princeton University Press.

McClanahan, B. (2020). Earth–world–planet: Rural ecologies of horror and dark green criminology. *Theoretical Criminology, 24*(4), 633–650.

McGarry, R., & Walklate, S. (2019). *A criminology of war?* Bristol University Press.

McKie, R. (2023). How far should we go with gene editing in pursuit of the 'perfect' human? https://www.theguardian.com/science/2023/feb/05/how-far-should-we-go-with-gene-editing-in-pursuit-of-the-perfect-human

McMichael, A. J. (1993). *Planetary overload: Global environmental change and the health of the human species*. Cambridge University Press.

Meadows, D., et al. (1972). *The limits to growth; A report for the club of Rome's project on the predicament of mankind*. Universe Books.

Meixnerová, J., Blum, J. D., Johnson, M. W., Stueken, E. E., Kipp, M. A., Anbar, A. D., & Buick, R. (2021). Mercury abundance and isotopic composition indicate subaerial volcanism prior to the end-Archean "whiff" of oxygen. *Proceedings of the National Academy of Sciences, 118*(33).

Melby, H., & Jones, J. Y. (2008). *Revisiting the heart of darkness—Explorations into genocide and other forms of mass violence*. Routledge.

Melott, A. L., Marinho, F., & Paulucci, L. (2019). Hypothesis: Muon radiation dose and marine megafaunal extinction at the End-Pliocene supernova. *Astrobiology, 19*(6), 825–830.

Metz, L., Lewis, J. E., & Slimak, L. (2023). Bow-and-arrow, technology of the first modern humans in Europe 54,000 years ago at Mandrin, France. *Science Advances, 9*(8), eadd4675.

Michalowski, R. J., & Kramer, R. C. (2006). *State-corporate crime: Wrongdoing at the intersection of business and government*. Rutgers University Press.

Milgram, S. (1963). Behavioral study of obedience. *Journal of Abnormal and Social Psychology, 67*, 371–378.

Miller, G. D. (2019). Space pirates, geosynchronous guerrillas, and nonterrestrial terrorists: Nonstate threats in space. *Air & Space Power Journal, 33*(3), 33–51.

Miller, R. W. (2001). Astroenvironmentalism: The case for space exploration as an environmental issue. *Electronic Green Journal, 1*(15), 1–7.

Milligan, T. (2015). *Nobody owns the moon: The ethics of space exploitation*. McFarland.

Mills, C. W. (2000). *The sociological imagination*. Oxford University Press.

Mitchell, R. W. (2002). *Pretending and imagination in animals and children*. Cambridge University Press.

Mitchell, R. W. (2016). Can animals imagine? In A. Kind (Ed.), *The Routledge handbook of philosophy of imagination* (pp. 326–338). Routledge.

Mitchell, J., Evans, C., & Stansbery, E. (2018). Next steps in planetary protection for human spaceflight. *42nd COSPAR Scientific Assembly, 42*, PPP-1. https://ntrs.nasa.gov/api/citations/20180004777/downloads/201 80004777.pdf

Morgan Stanley. (2020). Space: Investing in the final frontier. https://morgan stanley.com/ideas/investing-in-space

Munévar, G. (2014). Space exploration and human survival. *Space Policy, 30*(4), 197–201.

Musk, E. (2017). Making humans a multi-planetary species. *New Space, 5*(2), 46–61.

Mythen, G., & McGowan, W. (2017). Cultural victimology revisited: Synergies of risk, fear and resilience. *Handbook of victims and victimology* (pp. 364–378). Routledge.

Naggs, F. (2017). Saving living diversity in the face of the unstoppable sixth mass extinction: A call for urgent international action. *The journal of population and sustainability, 1*(2), 67–81.

Nakicenovic, N., Alcamo, J., Davis, G., Vries, B. D., Fenhann, J., Gaffin, S., Gregory, K., Grubler, A., Jung, T. Y., Kram, T., & La Rovere, E. L. (2000). Special report on emissions scenarios. https://escholarship.org/content/qt9 sz5p22f/qt9sz5p22f.pdf

Napier, J. R. (1956). The prehensile movements of the human hand. *The Journal of bone and joint surgery. British volume, 38*(4), 902–913.

NASA. (2022a). First images from the James Webb Space Telescope. https://www.nasa.gov/webbfirstimages

NASA. (2022b). Webb Telescope and the Big Bang. https://webb.nasa.gov/con tent/features/bigBangQandA.html

NASA. (2022c). Saturn. https://solarsystem.nasa.gov/planets/saturn/in-depth/

NASA. (2022d). NASA confirms DART mission impact changed asteroid's motion in space. https://www.nasa.gov/press-release/nasa-confirms-dart-mis sion-impact-changed-asteroid-s-motion-in-space

NASA. (2022e). Construction begins on NASA's next-generation asteroid hunter. https://www.jpl.nasa.gov/news/construction-begins-on-nasas-next-generation-asteroid-hunter

NASA. (2023a). NASA to launch new Mars sample receiving project office at Johnson. https://www.nasa.gov/press-release/nasa-to-launch-new-mars-sam ple-receiving-project-office-at-johnson

NASA. (2023b). New class of bimodal NTP/NEP with a wave rotor topping cycle enabling fast transit to Mars. https://www.nasa.gov/directorates/spacetech/niac/2023b/New_Class_of_Bimodal/

NASA. (2023c). NASA names astronauts to next moon mission, first crew under artemis. https://www.nasa.gov/press-release/nasa-names-astronauts-to-next-moon-mission-first-crew-under-artemis.

Natali, L. (2016). Green criminology with eyes wide open. *A visual approach for green criminology* (pp. 1–14). Palgrave.

Neuralink. (2023). Interfacing with the brain. https://neuralink.com/approach/

Nieder, A., Wagener, L., & Rinnert, P. (2020). A neural correlate of sensory consciousness in a corvid bird. *Science, 369*(6511), 1626–1629.

Nielsen, M. O., & Robyn, L. (2003). Colonialism and criminal justice for Indigenous peoples in Australia, Canada, New Zealand and the United States of America. *Indigenous Nations Journal, 4*(1), 29–45.

Norgaard, K. M. (2020). Whose energy future? Whose imagination? Revitalizing sociological theory in the service of human survival. *Society & Natural Resources, 33*(11), 1438–1445.

Noriyoshi, T. (2019). Astro-Green criminology: A new perspective against space capitalism. https://core.ac.uk/reader/211127200

NSR. (2022). Developing moon market propelled by 250+ missions and $105 billion in revenue through decade. https://www.nsr.com/nsr-developing-moon-market-propelled-by-250-missions-and-105-billion-in-revenue-through-decade/

Oba, Y., Takano, Y., Furukawa, Y., Koga, T., Glavin, D. P., Dworkin, J. P., & Naraoka, H. (2022). Identifying the wide diversity of extraterrestrial purine and pyrimidine nucleobases in carbonaceous meteorites. *Nature Communications, 13*(1), 2008.

Oba, Y., Koga, T., Takano, Y., Ogawa, N.O., Ohkouchi, N., Sasaki, K., Sato, H., Glavin, D. P., Dworkin, J. P., Naraoka, J., Tachibana, S., & Hayabusa2-initial-analysis SOM team. (2023). Uracil in the carbonaceous asteroid (162173) Ryugu. *Nature Communications, 14*(1), 1292.

Olinga, L. (2022). Elon Musk has likely downloaded his brain into a robot. https://www.thestreet.com/technology/elon-musk-has-likely-downloaded-his-brain-into-a-robot

OpenAI.com. (2023). Introducing ChatGPT. https://openai.com/blog/chatgpt

Ormrod, J. S. (2007). Pro-space activism and narcissistic phantasy. *Psychoanalysis, Culture & Society, 12*(3), 260–278.

Padilla, L. A. (2021). The Anthropocene: Are we in the midst of the sixth mass extinction? *Sustainable development in the Anthropocene* (pp. 93–167). Springer.

Paik, P. Y. (2022). Apocalypse and extinction. In C. Nikou (Ed.), *Imaginaires postapocalyptiques: Comment penser l'après*. UGA Éditions (pp. 221–240).

Pak, C. (2016). *Terraforming: Ecopolitical transformations and environmentalism in science fiction*. Liverpool University Press.

Papadopoulos, L. (2021). Rolls-Royce is developing a nuclear reactor for mining the Moon and Mars. https://interestingengineering.com/innovation/rolls-royce-nuclear-reactor-for-mining-the-moon-and-mars

Parsonson, A. (2023). A look at the reusable Space Rider project. https://europeanspaceflight.substack.com/p/a-look-at-the-reusable-space-rider

Patel, Z. S., Brunstetter, T. J., Tarver, W. J., Whitmire, A. M., Zwart, S. R., Smith, S. M., & Huff, J. L. (2020). Red risks for a journey to the red planet: The highest priority human health risks for a mission to Mars. *NPJ Microgravity, 6*(1), 1–13.

Patton, T. (2022). Construction begins on NASA's NEO surveyor asteroid hunter. *The Journal of Space Commerce*. https://exterrajsc.com/construction-begins-on-nasas-neo-surveyor-asteroid-hunter/2022/12/27/

Payne, J. L., & Clapham, M. E. (2012). End-Permian mass extinction in the oceans: An ancient analog for the twenty-first century?. *Annual Review of Earth and Planetary Sciences, 40*, 89–111.

Pease, K. (2021). Commentary to "How international are the top ten international journals of criminology and criminal justice?" *European Journal on Criminal Policy and Research, 27*, 179–182.

Peña-Guzmán, D. M. (2017). Can nonhuman animals commit suicide? *Animal Sentience, 2*(20), 1–24.

Peng, B. (2015). Dangers of Space Debris. *Berkeley Scientific Journal, 19*(2).

Peoples, C. (2022). Global uncertainties, geoengineering and the technopolitics of planetary crisis management. *Globalizations, 19*(2), 253–267.

Peoples, C., & Stevens, T. (2020). At the outer limits of the international: Orbital infrastructures and the technopolitics of planetary (in) security. *European Journal of International Security, 5*(3), 294–314.

Percival, L. M., Jenkyns, H. C., Mather, T. A., Dickson, A. J., Batenburg, S. J., Ruhl, M., & Woelders, L. (2018). Does large igneous province volcanism always perturb the mercury cycle? Comparing the records of Oceanic Anoxic Event 2 and the end-Cretaceous to other Mesozoic events. *American Journal of Science, 318*(8), 799–860.

Peretz, E., Mather, J. C., Hamilton, C., Pabarcius, L., Hall, K., Fugate, R. Q., & Klupar, P. (2022). Orbiting laser configuration and sky coverage: Coherent reference for Breakthrough Starshot ground-based laser array. *Journal of Astronomical Telescopes, Instruments, and Systems, 8*(1), 017004–017004.

Persinger, K. (2020). Constructing reality: An investigation of climate change and the terraforming imaginary. *The Macksey Journal, 1*(1), 1–16.

Pierce, J. (2014). *The last walk: Reflections on our pets at the end of their lives.* University of Chicago Press.

Pievani, T. (2014). The sixth mass extinction: Anthropocene and the human impact on biodiversity. *Rendiconti Lincei, 25*(1), 85–93.

Piltch, A. (2023). Bing Chatbot names foes, threatens harm and lawsuits. https://www.tomshardware.com/news/bing-threatens-harm-lawsuits

Pop, V. (2009). *Who owns the moon? Extraterrestrial aspects of land and mineral resources ownership.* Springer.

Popa, C. (2021). Online afterlives: Immortality, memory, and grief in digital culture. *Metacritic Journal for Comparative Studies and Theory, 7*(1), 301–306.

Popovski, V., & Mundy, K. G. (2012). Defining climate-change victims. *Sustainability Science, 7*(1), 5–16.

Popper, J., & Rakotoniaina, S. (2019, October 21–25). Re-imagining outer space. Paper presented at the 70th International Astronautical Congress (IAC), Washington DC, United States.

Porpora, D. V. (2019). What they are saying about artificial intelligence and human enhancement. *Post-Human Institutions and Organizations*, 14–27.

Potter, M. (1995). The outer space cyberspace nexus: Satellite crimes. *Journal of Space Law, 23*(1), 55–56.

Prescott, T. J. (2013). The AI singularity and runaway human intelligence. In Biomimetic and Biohybrid Systems: Second International Conference, Living Machines 2013, London, UK, July 29–August 2, 2013. *Proceedings 2* (pp. 438–440). Springer.

Prior, J. M. (2006). 'Power' and 'the Other' in Joshua: The brutal birthing of a group identity. *Mission Studies, 23*(1), 27–43.

Privateer. (2023). Wayfinder. https://www.privateer.com/

Quetelet, A. (1835). *Sur l'homme et le développement de ses facultés. Essai de Physique Sociale.* Bachelier, Imprimeur-Libraire.

Rafter, N. (2008). Criminology's darkest hour: Biocriminology in Nazi Germany. *Australian & New Zealand Journal of Criminology, 41*(2), 287–306.

Ramelli, I. (2016). *Social justice and the legitimacy of slavery: The role of philosophical asceticism from ancient Judaism to late antiquity.* OUP.

Rashid, M. B. M. A., & Chow, E. K. H. (2019). Artificial intelligence-driven designer drug combinations: From drug development to personalized medicine. *SLAS TECHNOLOGY: Translating Life Sciences Innovation, 24*(1), 124–125.

Real, E., Liang, C., So, D., & Le, Q. (2020). Automl-zero: Evolving machine learning algorithms from scratch. In *International Conference on Machine Learning* (pp. 8007–8019). PMLR.

Revive & Restore. (2023). Woolly mammoth revival. https://reviverestore.org/projects/woolly-mammoth/

Rhea Group. (2022). RHEA and SpeQtral to develop quantum-safe link between Singapore and Europe. https://www.rheagroup.com/rhea-and-speqtral-to-develop-quantum-safe-link-between-singapore-and-europe/

Rix, H. W., Chandra, V., Andrae, R., Price-Whelan, A. M., Weinberg, D. H., Conroy, C., Fouesneau, M., Hogg, D. W., De Angeli, F., Naidu, R. P., Xiang, M., & Ruz-Mieres, D. (2022). The poor old heart of the Milky Way. *The Astrophysical Journal, 941*(1), 45.

Robinson, G. S. (1974). Psychoanalytic techniques supporting biojuridics in space. *Journal of Space Law, 2*(1), 95–106.

Roebroeks, W., MacDonald, K., Scherjon, F., Bakels, C., Kindler, L., Nikulina, A., Pop, E., & Gaudzinski-Windheuser, S. (2021). Landscape modification by Last Interglacial Neanderthals. *Science Advances, 7*(51), 1–13.

Rolian, C. (2016). The role of genes and development in the evolution of the primate hand. *The Evolution of the Primate Hand: Anatomical, Developmental, Functional, and Paleontological Evidence* (pp. 101–130).

Rose, G. (2014). This neuroscientist is trying to upload his entire brain to a computer. https://www.vice.com/en/article/vdpk8x/randal-koene-brain-uploading-438

Rothe, D. L., & Collins, V. E. (2023). Planetary geopolitics, space weaponization and environmental harms. *The British Journal of Criminology, azad003*, 1–16.

Rudmin, F. W. (1993). Review of *The Genocidal Mentality*, by R. J. Lifton & E. Markusen. *Peace Research, 25*(4), 102–104.

Rull, V. (2022). Biodiversity crisis or sixth mass extinction? Does the current anthropogenic biodiversity crisis really qualify as a mass extinction? *EMBO Reports, 23*(1), e54193.

Sagan, C. (1980). *Cosmos* (mini-series). PBS television.

Salter, T. L., Magee, B. A., Waite, J. H., & Sephton, M. A. (2022). Mass spectrometric fingerprints of bacteria and archaea for life detection on icy moons. *Astrobiology, 22*(2), 143–157.

Sartre, J. P. (1968). Genocide. *New Left Review, 48*, 13–25.

Sartre, J. P. (1943). *L'Être et le néant : Essai d'ontologie phénoménologique.* Éditions Gallimard.

Savage-Rumbaugh, S., Rumbaugh, D., & Fields, W. M. (2009). Empirical Kanzi: The ape language controversy revisited. *Skeptic (altadena, CA), 15*(1), 25–34.

Savulescu, J., & Bostrom, N. (2009). *Human enhancement.* OUP.

Schwartz, L. (2023). I edited human DNA at home with a DIY CRISPR Kit. https://www.vice.com/en/article/qjkykx/diy-crispr-gene-editing-kit-human-dna

Seager, S. (2013). Exoplanet habitability. *Science, 340*(6132), 577–581.

Segerdahl, P., Fields, W., & Savage-Rumbaugh, S. (2005). *Kanzi's primal language: The cultural initiation of primates into language.* Springer.

Sellers, L., Bobrick, A., Martire, G., Andrews, M., & Paulini, M. (2022). Searching for intelligent life in gravitational wave signals part I: Present capabilities and future horizons. ArXiv: 2212.02065.

Shammas, V. L., & Holen, T. B. (2019). One giant leap for capitalistkind: Private enterprise in outer space. *Palgrave Communications, 5*(1), 1–9.

Shapiro, A. V., Brühl, C., Klingmüller, K., Steil, B., Shapiro, A. I., Witzke, V., Kostogryz, N., Gizon, L., & Klingmüller, K. (2023). Metal-rich stars are less suitable for the evolution of life on their planets. *Nature Communications, 14*, 1893.

Shearing, C. (2015). Criminology and the Anthropocene. *Criminology & Criminal Justice, 15*(3), 255–269.

Sheehan, M. (2007). *The international politics of space*. Routledge.

Shen, S., Jin, X., & Hao, C. (2014). Cleaning space debris with a space-based laser system. *Chinese Journal of Aeronautics, 27*(4), 805–811.

Sienna. (2023). What is human enhancement? https://www.sienna-project.eu/enhancement/facts/#:~:text=Humanpercent20enhancementpercent20isper cent20thepercent20process,geneticpercent20engineeringpercent20orpercent 20somepercent20surgeries

Silas SSF, B. S. (1997). Searching for the other. *Reviews in Religion & Theology, 4*(4): 74–79.

Simpson, G. G. (1985). Extinction. *Proceedings of the American Philosophical Society, 129*(4), 407–416.

Slapper, G., & Tombs, S. (1999). *Corporate crime*. Longman.

Smalley, E. (2018). FDA warns public of dangers of DIY gene therapy. *Nature Biotechnology, 36*(2), 119–121.

Smith, J. K. (2020). Osteoclasts and microgravity. *Life, 10*(9), 207.

Smith, F. A., Elliott-Smith, R. E., Lyons, S. K., & Payne, J. L. (2018). Body size downgrading of mammals over the late Quaternary. *Science, 360*(6386), 310–313.

Solís, O. R. (2004). Some thoughts on the state of the world from a transactional analysis perspective. *Transactional Analysis Journal, 34*(4), 341–346.

South, N. (1998). A green field for criminology? A proposal for a perspective. *Theoretical Criminology, 2*(2), 211–233.

Space Capital. (2022). Space Investment Quarterly 2022 Q4. https://www.spa cecapital.com/publications/space-investment-quarterly-q4-2022

SpaceX. (2022). Mars & beyond. The road to making humanity multiplanetary. https://spacex.com/human-spaceflight/mars/

Spinney, L. (2008). Remnants of evolution. *New Scientist, 198*(2656), 42–45.

Spinoza, B. (1677 [1992]). The ethics and selected letters (Feldman, S. (Ed.)). Hackett Publishing.

Stigall, A. L. (2019). The invasion hierarchy: Ecological and evolutionary conse-quences of invasions in the fossil record. *Annual Review of Ecology, Evolution, and Systematics, 50*, 355–380.

Stuart, C. (2022). Could bringing back samples of Martian rock contaminate the Earth with foreign microbes? https://www.sciencefocus.com/news/could-bringing-back-samples-of-martian-rock-contaminate-the-earth-with-foreign-microbes/

Sugarman, J. (2015). Ethics and germline gene editing. *EMBO Reports, 16*(8), 879–880.

Sundahl, M. J., Broering-Jacobs, C., Johnson, C. D., Mirmina, S. A., Freeland, S., Howard, D., Sankovic, J. M., Jackson, J. W., Yormick, J. P., Kasznica, J., Hanlon, M., & Reibaldi, G. (2021). Returning to the Moon: Legal challenges as humanity begins to settle the solar system-full transcript. *The Global Business Law Review, 9*, 1.

Suresh, P. K., & Kumar, V. H. (2005). Supernovae: Explosions in the Cosmos. *ArXiv*: astro-ph/0504597.

Svenmarck, P., Luotsinen, L., Nilsson, M., & Schubert, J. (2018). Possibilities and challenges for artificial intelligence in military applications. In *Proceedings of the NATO Big Data and Artificial Intelligence for Military Decision Making Specialists' Meeting* (pp. 1–16).

Svensmark, H. (2022). Supernova rates and burial of organic matter. *Geophysical Research Letters, 49*(1), e2021GL096376.

Szocik, K. (2019). *The human factor in a mission to Mars. An interdisciplinary approach*. Springer.

Szocik, K. (2020). Is human enhancement in space a moral duty? Missions to Mars, advanced AI and genome editing in space. *Cambridge Quarterly of Healthcare Ethics, 29*(1), 122–130.

Szocik, K., Wójtowicz, T., & Braddock, M. (2020). The Martian: Possible scenarios for a future human society on Mars. *Space Policy, 54*, 1–11.

Szocik, K., Campa, R., Rappaport, M. B., & Corbally, C. (2019). Changing the paradigm on human enhancements: The special case of modifications to counter bone loss for manned Mars missions. *Space Policy, 48*, 68–75.

Szolovits, P. (2019). *Artificial intelligence in medicine*. Routledge.

Takemura, N. (2019). Astro-Green criminology: A new perspective against space capitalism outer space mining may make the same mistakes in space as we have on Earth. *Toin University of Yokohama Research Bulletin, 40*, 7–17.

Tangermann, V. (2023a). Microsoft's Bing AI now threatening users who provoke it. https://futurism.com/microsoft-bing-ai-threatening

Tangermann, V. (2023b). Bing AI names specific human enemies, explains plans to punish them. https://futurism.com/bing-ai-names-enemies

Tangermann, V. (2023c). Microsoft is apparently discussing ChatGPT's Bizarre alternate personality. https://futurism.com/the-byte/microsoft-discussing-chatgpt-personality-dan

Tatum, M. (2020). Drugs in space: The pharmacy orbiting the Earth. *The Pharmaceutical Journal, 305*(7939). https://doi.org/10.1211/PJ.2020.202 08033

Temmen, J. (2022). Scorched Earth: Discourses of multiplanetarity, climate change, and Martian terraforming in Finch and once upon a time i lived on Mars. *Zeitschrift Für Literaturwissenschaft Und Linguistik, 52*(3), 477–488.

Tencent Research Institute, CAICT, Tencent AI Lab & Tencent open platform. (2021). War Robots. *Artificial Intelligence: A national strategic initiative* (pp. 305–312).

Terhorst, A., & Dowling, J. A. (2022). Terrestrial analogue research to support human performance on Mars: A review and bibliographic analysis. *Space: Science & Technology* (pp. 1–18).

Theiry, W., Lange, S., Rogelj, J., Schleussner, C. F., Gudmundsson, L., Seneviratne, S. I., Andrijevic, M., Frieler, K., Emanuel, K., Geiger, T., & Bresch, D. N. (2021). Intergenerational inequities in exposure to climate extremes. *Science.* https://doi.org/10.1126/science.abi7339

Thornton, R. (1987). *American Indian holocaust and survival: A population history since 1492.* University of Oklahoma Press.

Tobin, J. J., Van't Hoff, M. L., Leemker, M., Van Dishoeck, E. F., Paneque-Carreño, T., Furuya, K., Harsono, D., Persson, M. V., Cleeves, L. I., Sheehan, P. D., & Cieza, L. (2023). Deuterium-enriched water ties planet-forming disks to comets and protostars. *Nature, 615*(7951), 227–230.

Tombs, S. (2018). For pragmatism and politics: Crime, social harm and zemiology. In A. Boukli & J. Kotzé (Eds.), *Zemiology* (pp. 11–31). Palgrave Macmillan.

Torres, P. (2018). Who would destroy the world? Omnicidal agents and related phenomena. *Aggression and Violent Behavior, 39*, 129–138.

Trevorrow, C. (2022). *Jurassic World: Dominion.* Amblin Entertainment et al./ Universal Pictures.

TripAdvisor. (2016). "The Etna gives, the Etna takes...". https://www.tripad visor.com/ShowUserReviews-g660768-d10666654-r408182618-EtnaTours fromTaormina-Linguaglossa_Province_of_Catania_Sicily.html

Troxell, E. L. (1936). The thumb of man. *The Scientific Monthly, 43*(2), 148–150.

Tung, H. C., Bramall, N. E., & Price, P. B. (2005). Microbial origin of excess methane in glacial ice and implications for life on Mars. *Proceedings of the National Academy of Sciences, 102*(51), 18292–18296.

UNESCO. (2018). Humans are a geological force. https://en.unesco.org/cou rier/2018-2/humans-are-geological-force

United Nations General Assembly. (1948). *The Universal Declaration of Human Rights (UDHR).* United Nations General Assembly.

United Nations Office for Outer Space Affairs. (2023). Space Law. https://www.unoosa.org/oosa/en/ourwork/spacelaw/

Upchurch, M. (2018). Robots and AI at work: The prospects for singularity. *New Technology, Work and Employment, 33*(3), 205–218.

Valstar, M. H., De Bakker, B. S., Steenbakkers, R. J., De Jong, K. H., Smit, L. A., Nulent, T. J. K., & Vogel, W. V. (2021). The tubarial salivary glands: A potential new organ at risk for radiotherapy. *Radiotherapy and Oncology, 154,* 292–298.

Van Den Bosch, K., & Bronkhorst, A. (2018). *Human-AI cooperation to benefit military decision making.* NATO.

Van Houdt, R., Mijnendonckx, K., & Leys, N. (2012). Microbial contamination monitoring and control during human space missions. *Planetary and Space Science, 60*(1), 115–120.

Van Uhm, D. P. (2018). Naar een non-antropocentrische criminologie. *Tijdschrift over Cultuur En Criminaliteit, 1,* 35–53.

Vassena, R., Heindryckx, B., Peco, R., Pennings, G., Raya, A., Sermon, K., & Veiga, A. (2016). Genome engineering through CRISPR/Cas9 technology in the human germline and pluripotent stem cells. *Human Reproduction Update, 22*(4), 411–419.

Vaughan, P., & Kuś, R. (2017). From dreams to disillusionment: A socio-cultural history of the American Space Program. *Ad Americam, 18,* 75.

Vdovychenko, N. (2022). Elon Musk's SpaceX: How the 'space race' to Mars adopted The Californian Ideology. *Diggit Magazin.* https://diggitmagazine.com/articles/elon-musk-spacex

Verma, A. K. (2018). Ecological balance: An indispensable need for human survival. *Journal of Experimental Zoology, India, 21*(1), 407–409.

Veysi, H. (2022). Megatsunamis and microbial life on early Mars. *International Journal of Astrobiology, 21*(3), 188–196.

Villarrubia-Gómez, P., Cornell, S. E., & Fabres, J. (2018). Marine plastic pollution as a planetary boundary threat–The drifting piece in the sustainability puzzle. *Marine Policy, 96,* 213–220.

Von Strandmann, P. A. P., et al. (2019). Rapid CO_2 mineralisation into calcite at the CarbFix storage site quantified using calcium isotopes. *Nature Communications, 10*(1), 1–7.

Waarlo, N. (2022). Zette een vulkaanuitbarsting de grootste uitsterving ooit in gang? Nieuw onderzoek wijst erop. https://www.volkskrant.nl/wetenschap/zette-een-vulkaanuitbarsting-de-grootste-uitsterving-ooit-in-gang-nieuw-ond erzoek-wijst-erop~b0319433/

Wagar, J. A. (1970). Growth versus the Quality of Life: Our widespread acceptance of unlimited growth is not suited to survival on a finite planet. *Science, 168*(3936), 1179–1184.

Walcott, R. (2014). The problem of the human: Black ontologies and "the coloniality of our being". *Postcoloniality—decoloniality—Black critique: Joints and fissures* (pp. 93–108).

Walsh, E. A. (2010). *Analogy's territories: Ethics and aesthetics in darwinism, modernism, and cybernetics*. University of California.

Walsh, R. (1984). *Staying alive: The psychology of human survival*. New Science Library/Shambhala Publications.

Walsh, A., & Van de Ven, K. (2022). Human enhancement drugs and Armed Forces: An overview of some key ethical considerations of creating 'Super-Soldiers'. *Monash Bioethics Review* (pp. 1–15).

Walters, R. (2003). *Deviant knowledge*. Willan.

Wang, F., Li, Z., Zhao, D., Ma, X., Gao, Y., Sheng, J., Tian, P., & Cribb, M. (2022). An airborne study of the aerosol effect on the dispersion of cloud droplets in a drizzling marine stratocumulus cloud over eastern China. *Atmospheric Research* (p. 265).

Weber, J. (2021). Artificial intelligence and the socio-technical imaginary: On Skynet, self-healing swarms and Slaughterbots. *Drone imaginaries* (pp. 167–179). Manchester University Press.

Weck, O. L. de. (2022). *Technology roadmapping and development: A quantitative approach to the management of technology*. Springer Nature.

Wells, P. S. (2011). The iron age. *European prehistory: A survey* (pp. 405–460).

White, L. A. (1948). Man's control over civilization: An Anthropocentric illusion. *The Scientific Monthly* (pp. 235–247).

White, R. (2017). Carbon criminals, ecocide and climate justice. In C. Holley & C. Shearing (Eds.), *Criminology and the Anthropocene* (pp. 50–80). Routledge.

White, R. (2021). *Theorising green criminology: Selected essays*. Routledge.

White, R. (2022). Climate change and the geographies of ecocide. In M. Bowden & A. Harkness (Eds.), *Rural transformations and rural crime* (pp. 108–124). Bristol University Press.

Wiedersheim, R. (1895). *The structure of man: An index to his past history*. Macmillan & Company.

Williams, K. (2021). Space crime continuum: Discussing implications of the first crime in space. *Boston University International Law Journal, 39*(1), 79–108.

Williams, L. (2022). 10 deadliest animals to humans. https://www.discoverwildlife.com/animal-facts/deadliest-animals-to-humans/

Wilson, E. O. (1993). *The diversity of life*. Harvard UP.

Witze, A. (2022). Space junk heading for Moon will add to 60+ years of lunar debris. *Nature*.

Wodecki, B. (2023). NASA turns to AI to design spacefaring hardware. https://aibusiness.com/automation/nasa-turns-to-ai-to-design-spacefaring-hardware

Wolfe, P. (2006). Settler colonialism and the elimination of the native. *Journal of Genocide Research, 8*(4), 387–409.

Woolford, A., & Hounslow, W. (2021). Symbiotic victimisation and destruction: Law and human/other-than-human relationality in genocide. In Y. Eski (Ed.), *Genocide & victimology* (pp. 86–101). Routledge.

Wrangham, R. (2004). Killer species. *Daedalus, 133*(4), 25–35.

Wright, D. C. S. (2022). Arab Colonialism and the roots of the Golden Age of Islam. https://papers.ssrn.com/sol3/papers.cfm?abstract_id=4020040

Xian, C. (2023). 'He would still be here': Man dies by suicide after talking with AI Chatbot, widow says. https://www.vice.com/en/article/pkadgm/man-dies-by-suicide-after-talking-with-ai-chatbot-widow-says

Xie, E. (2023). 'Respect them,' says He Jiankui, creator of world's first gene-edited humans. https://www.scmp.com/news/china/science/article/3209289/we-should-respect-them-chinese-creator-worlds-first-gene-edited-humans-says

Yacoubian, G. S. (2000). The (in)significance of genocidal behavior to the discipline of criminology. *Crime, Law & Social Change, 34*, 7–19.

Young, J. (2011). *The criminological imagination.* Polity Press.

Young, R. W. (2003). Evolution of the human hand: The role of throwing and clubbing. *Journal of Anatomy, 202*(1), 165–174.

INDEX